LINEAR ALGEBRA WITH MAPLE®

WILLIAM C. BAULDRY
APPALACHIAN STATE UNIVERSITY

BENNY EVANS
OKLAHOMA STATE UNIVERSITY

JERRY JOHNSON
UNIVERSITY OF NEVADA, RENO

JOHN WILEY & SONS, INC.
New York   Chichester   Brisbane   Toronto   Singapore

ISBN 0-471-06368-1

Printed in the United States of America

10 9 8 7 6 5 4 3 2 1

Printed and bound by Malloy Lithographing, Inc.

TABLE OF CONTENTS

PREFACE

This manual is a translation to Maple V* of Benny Evans and Jerry Johnson's *Linear Algebra with DERIVE*®. The Maple V code was written by William Bauldry. Converting from one software package to another is never easy and is often quite difficult. Nevertheless, many items transfer almost directly to Maple from *DERIVE*. In the places where the significant differences between these computer mathematics systems made the text appear quite different, every attempt was made to keep the spirit of the original work.

This is an enrichment supplement to a traditional introductory linear algebra course. Its purpose is to help students (and teachers) use the Maple V program as a tool to solve problems that arise in such a course. This book adds a new dimension to the conventional linear algebra class by providing problems that go beyond the level of rote calculations and template exercises, but at the same time fit comfortably into the traditional course.

Here's an example: Describe all matrices that commute with a given 3 by 3 matrix. (See Exercise 7 in Chapter 3.) This exercise makes the student *use* the concept of commutativity, which leads to a system of nine equations in nine unknowns. This is a daunting task to solve with nothing but pencil and paper, but Maple will do it almost instantly. The student must then use the solution set in order to describe the required set of matrices. This exercise is typical of many in this text in that students are not asked simply to do a calculation with Maple, but they must actually solve a problem.

Key Features

- No prior knowledge of Maple is required. Appendix A summarizes important commands and will help novice users get started. Students with a little Maple experience should be able to begin in Chapter 1.

- Most problems go beyond the level of rote calculations and "template" exercises. The exercise sets start with examples that are similar to the solved problems and progress to more challenging problems.

---

*Maple is a registered trademark of Waterloo Maple Software, Inc., Waterloo, Ontario, Canada. For information about Maple contact Waterloo Maple Software at (519) 747-2373.

- Numerous applications are provided, including optimization, Markov chains, systems of differential equations, the Gauss-Seidel method, generalized inverses, curve fitting, rotation of axes, and cryptography.

- The *LU* and *QR* factorizations are presented.

- Many of the recommendations of the Linear Algebra Curriculum Study Group are followed.

- Optional code for automating several procedures is provided in Appendix B.

- Many of the problems have been class tested by the authors and others.

1. Structure of the Book

This manual is written to be used with any linear algebra text, but we were guided by the contents, terminology, and general flow of Howard Anton's *Elementary Linear Algebra*[†]. Each chapter has five parts: Introduction, Solved Problems, Exercises, Exploration and Discovery, and Laboratory Exercises. The **Introduction** is a brief statement of the nature of the chapter. **Solved Problems** are examples that provide a context for Maple instructions that one is likely to encounter in the chapter and may also contain useful mathematics hints as well. **Exercises** are problems that generally ask for well-defined answers. They include unusual problems, such as the example previously mentioned, as well as variations on conventional exercises that are sufficiently complex that solving them without assistance from a computer is not practical. **Exploration and Discovery** problems are those that go a bit beyond the conventional exercise. Some are more involved and difficult. Others may not simply ask for an answer, but invite one to explore a situation and then to discuss observations and findings. **Laboratory Exercises** are problems with structured responses suitable for use as a lab assignment. They can be detached and handed in.

Most Maple instructions are provided in the context of problems to be solved, but additional help is provided in Appendix A. Appendix B contains optional code for Maple functions that students or instructors will find useful.

The appendices should not be considered a substitute for the Maple *Library Reference Manual*. We have tried to include enough Maple instructions and suggestions in both the solved problems and Appendix A that the reader will not have to spend

---

[†]Howard Anton, *Elementary Linear Algebra*, 7th edition, John Wiley & Sons, Inc., N.Y., 1994.

a lot of time referring to the user manual, but ultimately the reader should consult it for details.

2. The Maple Program

> *The novice who has never used Maple before should begin by reading Appendix A.*

Maple is a comprehensive computer mathematics system that is available on over 20 platforms from micro- to supercomputers. Maple runs on all Macintosh computers having 4 or more megabytes of RAM, on PC-compatible computers (at least 386 class with 4 or more megabytes of extended RAM) under MicroSoft Windows or DOS, and on Unix and VMS systems.

Any sophisticated software takes some practice and experience to master, but we have used computer mathematics systems—*DERIVE* for Evans and Johnson, Maple and *DERIVE* for Bauldry—in our courses since their first release on microcomputers and are convinced that one of its strongest educational advantages is that it is easy for students to learn and to use. Nevertheless, Maple is powerful enough for professional applications and research in mathematics, science, and engineering.

This manual was written using Maple V Release 3. If you are using an earlier version, you will notice a few differences. We have tried to mention them on the rare occasions when they might cause confusion. The most conspicuous new feature beginning with Release 2 is true mathematics output. See Section A.2 in Appendix A for samples.

An important item to be alert to is the following: Maple distinguishes between the vector [2, 3] and the 1 by 2 matrix [2 3]. In Maple, a matrix is a two dimensional object—rows and columns—while a vector has only one dimension. For example, Maple will take the transpose of a matrix returning the answer you expect, but, since a vector has no rows and columns to interchange, the transpose of a vector returns an unevaluated function which cannot be used for further calculations! (The syntax for the transpose of A is `transpose(A)`.) Maple interprets vectors properly for multiplication, and hence, will compute A times [2, 3] correctly. For examples and further discussion, see Section A.4 in Appendix A.

3. Suggestions for Incorporating Computer Mathematics: What Has Worked for Us

- We normally spend a period with our classes at the computer laboratory in the first week or two of the semester to introduce them to computer mathematics.

(Our experience is that students quickly become relatively comfortable with the systems and that only minimal help is needed later.)

- After this initial session we may return to the lab with the class four or six more times during the semester. It is very important to integrate the lab experience with material that is currently being covered in the course. We often begin such a lab session by asking students to work through a solved problem and later give them a quiz that consists of a similar exercise. Grades in the lab are important. Even the best students appreciate credit for their work, and there are always students who will remain passive in the lab without the incentive of a grade.

  Quizzes and tests are also given in the lab. One of the course goals is for students to learn to decide which tool, the computer or hand calculation, is more appropriate for a given problem.

- We give outside lab assignments. The frequency may vary from five to ten assignments a semester. We require students to work together on the assignments and write up the results in their own words.

- We have found that some classroom discussion of an assigned exercise is desirable, including a few words about new computer commands. However, all a student needs should be in the **Solved Problems** and Appendix A.

- To avoid extra expense to the student, a school's lab might buy enough manuals for one class (say 30 or 40) and check them out to students during lab time.

4. Important Conventions

You will notice early in this book the issue of uppercase versus lowercase letters; Maple is case sensitive. We encourage you to read Section A.2 of Appendix A.

We have tried to follow three notational conventions in this manual.

1. When a key is to be depressed on the keyboard, it is put in a box. For example, $\boxed{enter}$ means you are to press the *enter* key. Sometimes you may have to hold down two keys at once. In this case, both keys will appear in the same box with a space between them; for example, $\boxed{Ctrl\ C}$ means you hold down the key marked *Ctrl* while you press the *C* key.

2. When you must enter an expression into Maple, it will be put in what is some-times called "typewriter" style. In this case, you type exactly the symbols you see except the period at the end. For example, if we want you to enter $\frac{x^2}{3}$, we will say "enter x^2/3". To have you enter the vector $(2,3)$, we may say "enter the vector([2,3])".

3. Maple sessions have no line numbers; this presents a problem when referring to figures in the text. We will refer to each Maple input-output pair as a *line* with a number corresponding to its order in the figure. Each line in a figure begins with a prompt symbol ">". The figures are copies of Maple sessions that are short enough that this should pose no difficulty.

5. A Word to Students

Maple has essentially automated most of the standard algebraic and matrix cal-culations you will encounter, just as scientific calculators have done with arithmetic. It will simplify complicated expressions, solve equations, draw graphs, and much more. It will also row-reduce matrices and find their determinants and inverses. But that doesn't mean linear algebra is obsolete or unimportant. Even though calculators will do arithmetic, we still have to know what questions to ask, understand what the answers mean, and realize when an obvious error has been made. In the same way, we still have to understand the definitions, concepts, and processes that are involved in linear algebra so that we will know what to tell Maple to do, understand what its answers mean, and be able to detect errors. Maple only does the *calculations*; you must do the *thinking*.

Always view any computer or calculator output critically. Be alert for answers that seem strange; you might have hit the wrong key, entered the wrong data, or made some other mistake. It is even possible that the program has a bug! If a problem asks for the cost of materials to make a shoe box and you get $123.28 or [4, 9] you should suspect something is wrong!

Clear communication is as important in mathematics as in other fields. You should always write your answers neatly in complete, logical sentences. Re-read what you have written and ask yourself whether it really makes sense.

Before you turn on the computer, work through as much of an assigned problem as you can with pencil and paper, taking note of exactly where you think the computer will be required and for what purpose. You may be surprised at how little time you will actually have to spend in front of the machine if you follow this advice.

With few exceptions, the problems in the section entitled **Exercises** are traditional in that they ask for a clearly defined answer; however, in the **Exploration and Discovery** section this may not be the case. Here we may not simply ask for an answer, but invite you to explore a situation and then to discuss your observations and findings. The instructions may even be vague on occasion. That's what discovery is all about. In fact, if your interpretation and analysis are correct, but not what we or your teacher expected, there is no cause for concern. On the contrary, it is a sign that you are indeed discovering mathematics.

6.   A Word to Instructors

Each chapter has a section entitled **Solved Problems** that may contain a straightforward example to provide a "warm-up problem" or a context for illustrating Maple commands and hints that may be useful later. However, we have largely avoided exercises that just show off Maple's power with no apparent mathematical lesson (find the inverse of the 10 by 10 Hilbert matrix $\{\frac{1}{i+j-1}\}_{i,j=1}^{10}$) or routine exercises that can be done easily by hand (row-reduce $\begin{bmatrix} 1 & 2 \\ -2 & -4 \end{bmatrix}$). Our opinion is that the computer should be a tool, not a crutch. To ask students to appeal to Maple for simple problems sends the wrong message, much like suggesting they use a calculator to divide 4 by 2.

Some of the problems in the **Exploration and Discovery** section do not simply ask for an answer, but invite students to explore and then discuss their observations. In some cases, we have deliberately avoided telling the students exactly what is expected of them. You may supply hints, guidance, or suggestions as you see fit.

We have used real numbers only in the exercises in this book, but Maple works in the complex field. It is quite easy to include complex numbers in some examples if you so desire. Read Section A.13 in Appendix A for information on handling complex numbers with Maple.

Finally, we have included Maple code in Appendix B that does specialized calculations, such as LU and QR factorizations. These functions might be a bit tedious for most students to enter, so you may want to do it yourself, save them for future use, and supply the files to your students.

7.   Acknowledgments

We would like to thank Professor Kent Nagle of the University of South Florida for class testing early versions of this material using Maple and for providing corrections and student reactions. Also, we wish to thank Stan Devitt of Waterloo Maple

Software for providing Maple V Release 3, through the author support program, and for TEXnical help in writing this book. Appreciation also goes to Jeffry Hirst of Appalachian State University for TEXnical help, often requested at the spur of the moment.

Chapter 1
SYSTEMS OF EQUATIONS

LINEAR ALGEBRA CONCEPTS

- **Solutions of systems of equations**

1.1 Introduction

Elementary linear algebra commonly begins with a study of systems of linear equations. The arithmetic required to solve such systems is often quite formidable, especially if the system involves many variables and equations or if the coefficients are complicated. Sometimes there is a unique solution, but often there may be no solution at all or a whole *set* of solutions whose description must be determined. Maple can handle each of these three cases, allowing attention to be focused on the utility of systems of equations rather than on the tedious calculations involved in solving them. If you are unfamiliar with Maple, read Appendix A before proceeding.

1.2 Solved Problems

Solved Problem 1: Solve the following system of equations.

$$\begin{aligned} 3x + 4y - 7z &= 8 \\ 2x - 3y + 4z &= 2 \\ 4x + 2y - 3z &= 4 \end{aligned}$$

Solution: To solve a system of linear equations with Maple, enter the equations in a set separated by commas and enclosed by set brackets. In our example, we enter `eqs := {3*x+4*y-7*z=8, 2*x-3*y+4*z=2, 4*x+2*y-3*z=4}`, we ask Maple to solve with `solve(eqs)`, and we see the unique solution in Figure 1.1. The form "`eqs :=`" defines `eqs` as the name of the set. Don't forget the trailing semicolon! (See section A.2 in Appendix A.)

Maple hint: An alternative method for entering the system of equations is to enter and name each equation separately (say, as `eq1 := 3*x+4*y-7*z=8`, with `eq2` and `eq3` similarly defined) and then `solve({eq1, eq2, eq3})`.

```
> eqs := {3*x+4*y-7*z=8, 2*x-3*y+4*z=2, 4*x+2*y-3*z=4};
```
$$eqs := \{3x + 4y - 7z = 8, 2x - 3y + 4z = 2, 4x + 2y - 3z = 4\}$$
```
> solve(eqs);
```
$$\left\{ z = \frac{-20}{81}, \quad y = \frac{-110}{21}, \quad x = \frac{16}{21} \right\}$$

Figure 1.1: A system of equations with a unique solution

Solved Problem 2: Solve the following system of equations. The system has an infinite number of solutions; find a specific solution with $x > 10$.

$$3x + 4y + z = 8$$
$$2x - y + 3z = 2$$

Solution: Enter eqs := {3*x+4*y+z=8, 2*x-y+3*z=2} and then, as above, enter solve(eqs). Maple recognizes that there are more variables than equations and chooses which variables to solve for. (Maple selects an order for the variables that depends on the state of the workspace and varies considerably between sessions.) To solve for x and y, we use solve(eqs, {x,y}). Maple will present the solution as in Figure 1.2. This shows that there are infinitely many solutions—one for each value of z we choose.

When a system of equations has infinitely many solutions, they are expressed in terms of one or more variables called *parameters*. In this case, z itself can be used as the parameter, or we can introduce another symbol. Below we have expressed the solution in terms of s.

$$x = -\frac{13}{11}s + \frac{16}{11}$$
$$y = \frac{7}{11}s + \frac{10}{11}$$
$$z = s$$

Solutions are presented in this way to emphasize the fact that the parameter s can be assigned any value at all, and each assignment determines values for x, y, and z that comprise a solution of the system. To determine a specific solution of the system with $x > 10$, we want to choose a value of s that makes $\frac{16-13s}{11} > 10$. The solution of this inequality is $s < -\frac{94}{13}$. (Maple will do this, but it's easy to solve by hand.) Thus, *any* value of s less than $-\frac{94}{13}$ will do; we choose $s = -8$.

```
> eqs := {3*x+4*y+z=8, 2*x-y+3*z=2};
```

$$eqs := \{3x + 4y + z = 8, \quad 2x - y + 3z = 2\}$$

```
> solve(eqs);
```

$$\left\{ z = -\frac{11}{13}x + \frac{16}{13}, \; y = -\frac{7}{13}x + \frac{22}{13}, x = x \right\}$$

```
> solve(eqs, {x,y});
```

$$\left\{ x = -\frac{13}{11}z + \frac{16}{11}, \; y = \frac{7}{11}z + \frac{10}{11} \right\}$$

```
> subs(z=-8, '');
```

$$\left\{ x = \frac{120}{11}, \; y = \frac{-46}{11} \right\}$$

Figure 1.2: A system of equations with many solutions

To find our final answer, we must substitute -8 for z. Enter `subs(z=-8,'')`. Recall that double quotes refer to the previous result (cf. Appendix A). See the last line of Figure 1.2. Again, we stress that there are many correct answers to the second part of the problem.

Solved Problem 3: Solve the following system of equations.

$$\begin{aligned} 3x + 4y + z &= 8 \\ 2x - y + 3z &= 2 \\ 3x + 4y + z &= 8 \end{aligned}$$

Solution: Notice that the first two equations are the same as those in Solved Problem 2. Since the third equation is the same as the first, the solution of the system is the same as the one in Solved Problem 2; but as we shall see, Maple presents it a little differently. Enter `eqs := {3*x+4*y+z=8, 2*x-y+3*z=2, 3*x+4*y+z=8}` and ask Maple to solve. Note that the third equation is the same as the first, and so, it only appears once in the display of the set `eqs`.

Remark on notation: We will refer to each Maple input-output pair as a *line* with a number corresponding to its order in the figure; for example, line 2 of Figure 1.3 has input `solve(eqs)` and output a set of solutions. Each line begins with the prompt symbol ">".

```
> eqs := {3*x+4*y+z=8, 2*x-y+3*z=2, 3*x+4*y+z=8};
```

$$eqs := \{3x + 4y + z = 8, 2x - y + 3z = 2\}$$

```
> solve(eqs);
```

$$\left\{ x = -\frac{13}{7}y + \frac{22}{7}, z = \frac{11}{7}y - \frac{10}{7}, y = y \right\}$$

```
> solve(eqs, {x,y});
```

$$\left\{ x = -\frac{13}{11}z + \frac{16}{11}, y = \frac{7}{11}z + \frac{10}{11} \right\}$$

Figure 1.3: | Alternate form for the case of many solutions |

We *should* be able to make this solution look like the one for Solved Problem 2. The solution in line 2 of Figure 1.2 is presented with solve variables y and z, while line 2 of Figure 1.3 uses solve variables x and z. Thus, they still look different. To correct this, solve for x and y by using solve('', {x,y}). The results are in Figure 1.3. The two solutions now look the same.

When solving systems of equations, we often have choices to make for the solve variables. Different choices may make the answers *look* different, but they are, in fact, equivalent.

Solved Problem 4: ACE Milling Company has received an order for 1000 pounds of a special feed that contains 4% fat, 15% fiber, and 15% protein. They must mix the required feed using the ingredients they have on hand: wheat mids*, rice mill feed, cotton seed meal, soybean hulls, and alfalfa.

How should they mix the feed if the cost is to be a minimum? Give the amounts of each ingredient used and the total cost.

The following table gives the percentages of fat, fiber, and protein plus the cost per pound for each ingredient.

---

*When the wheat kernel is removed to produce flour, some parts of the whole wheat including the bran are often pressed into pellets known as "wheat mids" and sold as a feed ingredient.

	% fat	% fiber	% protein	cost per pound ($)
w = wheat mids	3	5	15	0.05
r = rice mill feed	6	30	5	0.02
c = cotton seed meal	3	10	40	0.10
s = soybean hulls	2	35	10	0.05
a = alfalfa	4	25	20	0.08

Solution: The first step is to write the constraints on the composition of the feed as a system of equations. Let w, r, c, s, and a denote the total number of pounds of each ingredient (indicated in the table) to be used. Four percent of the 1000 pounds of feed is to be fat. Thus, from the % fat column in the table we obtain the following equation.

$$3\%w + 6\%r + 3\%c + 2\%s + 4\%a = 40$$

Remark: We will enter percents into Maple as decimals.

In a similar fashion, we obtain the following equations for fiber and protein.

$$5\%w + 30\%r + 10\%c + 35\%s + 25\%a = 150$$
$$15\%w + 5\%r + 40\%c + 10\%s + 20\%a = 150$$

Finally, the total amount of feed is 1000 pounds, so that $w + r + c + s + a = 1000$.

Notice that there are four equations in five unknowns, so we do not expect a unique solution. Enter the system and solve it, naming the solution **soln** as in Figure 1.4. (Even though Maple wraps the long expression around to a new line on the screen, it's still a single input line.) We can use any four of the variables as solve variables. The solution shown in Figure 1.4 is obtained using a, c, r, and s as solve variables.

The next step is to define the cost function as done in line 3 of Figure 1.4 and then use **subs** to replace a by $12w - 5900$, c by $-5w + 2600$, r by $-5w + 2800$, and s by $-3w + 1500$ to obtain the cost as a function of w. We can do this easily by substituting the solution into the cost function with **subs(soln, cost)**.

The cost, $0.26w - 81$, is clearly minimal if w is chosen as small as possible. But the physical constraints on the problem require that each variable a, c, r and s be nonnegative.

```
> eqs := { .03*w+.06*r+.03*c+.02*s+.04*a=40,
   .05*w+.3*r+.1*c+.35*s+.25*a=150,
   .15*w+.05*r+.4*c+.1*s+.2*a=150, w+r+c+s+a=1000 };
```

$$eqs := \{.15w + .05r + .4c + .1s + .2a = 150, .03w + .06r + .03c + .02s + .04a = 40,$$
$$.05w + .3r + .1c + .35s + .25a = 150, w + r + c + s + a = 1000\}$$

```
> soln := solve(eqs, {a,c,r,s});
```

$$soln := \{a = 12.w - 5900., r = -5.w + 2800, s = -3.w + 1500., c = -5.w + 2600\}$$

```
> cost := .05*w+.02*r+.1*c+.05*s+.08*a;
```

$$cost := .05w + .02r + .1c + .05s + .08a$$

```
> subs(soln, cost);
```

$$.26w - 81.00$$

```
> subs(w=1475/3, soln);
```

$$\{a = 0, \quad r = 341.666666, \quad s = 25.000000, \quad c = 141.666666\}$$

```
> subs(w=1475/3, cost);
```

$$46.833333$$

Figure 1.4: A feed ration of minimal cost

Combining **soln** from Figure 1.4 with these constraints gives the following inequalities.

$$
\begin{aligned}
12w - 5900 &\geq 0 \\
-5w + 2600 &\geq 0 \\
-5w + 2800 &\geq 0 \\
-3w + 1500 &\geq 0
\end{aligned}
$$

We want to find the *smallest* value of w that makes all four of these inequalities true. If we solve each inequality and compare the solutions, we find that $w = \frac{1475}{3}$. It is probably faster to do this by hand than it is to enter all the data into Maple.

Use **subs** to replace w by $\frac{1475}{3}$ in **soln** to obtain the amounts of the other ingredients as shown in Figure 1.4. If we put $w = \frac{1475}{3}$ into the cost function, we obtain the minimum cost of \$46.83 for the 1000 pounds of feed.

1.3   Exercises

1. Solve the following systems of equations. If the solution is not unique, find a specific solution with x between 20 and 30.

    (a)
$$\begin{aligned}
4x + 6y - z + 8w &= 5 \\
3x + 6y - 4z + w &= 8 \\
5x - y + 7z - 8w &= 3
\end{aligned}$$

    (b)
$$\begin{aligned}
6x - 2y + 8z + 4w &= 5 \\
5x - 4y - 3z + 2w &= 10 \\
2x - 7y - z - w &= -3 \\
x + 2y + 11z + 2w &= -5
\end{aligned}$$

    (c)
$$\begin{aligned}
2x + y + 3z - w &= 8 \\
x - y + 2z + 2w &= 4 \\
3x + 5z + w &= 12
\end{aligned}$$

    (d)
$$\begin{aligned}
x + y + z + w &= 8 \\
2x + 3y - z + w &= 4 \\
3x + 4y + 2w &= 5
\end{aligned}$$

2. Solve the following system of equations for x, y, and z.

$$\begin{aligned}
ax + 3y + 7z &= 9 \\
2x + 2y - z &= 4 \\
cx - y + 5z &= 2
\end{aligned}$$

    (a) What is the solution if $a = 3$ and $c = 7$?

    (b) What is the solution if $a = \frac{61}{9}$ and $c = 1$?

3. A bag contains 100 coins consisting of nickels, dimes and quarters. There is a total of $22.00 in the bag.

    (a) What is the maximum possible number of quarters?

    (b) What is the minimum possible number of quarters?

    (c) List all possibilities for the distribution of coins in the bag.

4. Suppose that *neither* of the following systems of equations has a solution.

$$
\begin{array}{rcl}
ax + 3y - z & = & 2 \\
bx + 2y - 3z & = & 5 \\
2x - y + 3z & = & 4
\end{array}
\qquad
\begin{array}{rcl}
ax - 2y + 7z & = & 9 \\
bx + 5y + 2z & = & 6 \\
4x + 7y - 8z & = & 1
\end{array}
$$

Find a and b.

5. The following problems are a continuation of Solved Problem 4.

(a) Find the maximum possible percentage of fat that the feed mixture can have if the other restrictions on the feed remain the same.

(b) Suppose that the required percentage of fat is 5%. Give the amounts of each ingredient so that the feed is of minimum cost and find this cost.

(c) Suppose that you find yourself overstocked with alfalfa. Rather than produce a mixture of minimal cost, you wish to fill the order using a maximum amount of alfalfa. Find the amounts of each ingredient and the total cost of this alfalfa-rich feed.

6. A manufacturer wishes to melt and combine some or all of six alloys that are available from a distributor to produce 100 pounds of a new alloy that is (by weight) 6% gold, 15% silver, and 30% iron. The five available alloys and their percentage content of gold, silver, and iron are given in the table below.

Alloy	% Gold	% Silver	% Iron
A	12	10	35
B	6	23	15
C	16	5	15
D	5	7	42
E	2	28	17
F	4	8	55

Prepare an order for the distributor that will allow you to produce the required alloy.

7. A grain silo typically has an irregularly shaped hopper at the base. At inventory time, someone stands at the top of the silo and drops a tape down the interior to the surface of the grain. The reading (in feet) is reported to the manager who consults a table (provided by the builder) that gives the number of cubic feet of grain remaining in the silo based on the distance from the surface of the grain to the top of the silo. An actual table for a silo owned by Stillwater Milling Company is as follows.

Depth	Feet^3	Depth	Feet^3	Depth	Feet^3	Depth	Feet^3
0	2815	13	1983	26	1151	39	319
1	2751	14	1919	27	1087	40	255
2	2687	15	1855	28	1023	41	196.5
3	2623	16	1791	29	959	42	147.5
4	2559	17	1727	30	895	43	106.9
5	2495	18	1663	31	831	44	74.9
6	2431	19	1599	32	767	45	49.5
7	2367	20	1535	33	703	46	30.5
8	2303	21	1471	34	639	47	16.5
9	2339	22	1407	35	575	48	7.2
10	2715	23	1343	36	511	49	1.8
11	2111	24	1279	37	447	50	0
12	2047	25	1215	38	383		

We want to find a function $C(x)$ that gives the cubic feet of grain in terms of the distance x from the surface of the grain to the top of the silo. The data points 0 through 40 correspond to the regular cylindrical portion of the silo, and thus are described by a linear function $f(x) = ax + b$ for a suitable choice of a and b. The last 10 data points for depths 41 through 50 cannot be accurately approximated by a linear function, so the idea is to find a polynomial function $g(x)$ that does. When we do, we get

$$C(x) = \begin{cases} f(x) & \text{if } x \leq 40 \\ g(x) & \text{if } 40 < x \leq 50 \end{cases}$$

(a) Find $f(x)$.

(b) There are a number of ways to find a function $g(x)$ that approximates the last 10 data points. (We shall return to this problem once more when we look at least-squares approximations.) We will use a fourth-degree polynomial that passes through the data points 41, 43, 45, 47, and 50, and check that the function doesn't miss the remaining points at depths 42, 44, 46, 48, and 49 feet by very much.

If $g(x) = ax^4 + bx^3 + cx^2 + dx + e$ fits the data points for depths 41, 43, 45, 47, and 50 feet, write a system of five equations in five unknowns that must be solved to determine the coefficients a, b, c, d, and e.

(c) Find $g(x)$.

(d) For depths 42, 44, 46, and 48 feet, calculate $g(x)$ and the error.

(e) Find the approximate amount of grain in the silo when the depth is 44.5 feet.

1.4   Exploration and Discovery

1. Ask Maple to solve the following system of equations for x and y.

$$
\begin{aligned}
ax + by &= c \\
dx + ey &= f
\end{aligned}
$$

Is this solution correct for all values of a, b, d, and e? If not, analyze the solution in cases where Maple's solution is not correct and discuss your findings.

2. Solve the following system of equations using x and y as solve variables.

$$
\begin{aligned}
x + y + 2z + 2w &= 3 \\
2x + y + 3z + 3w &= 4
\end{aligned}
$$

Ask Maple to solve the system once more using z and w as solve variables. Explain what happens. In general, which variables in a system of linear equations can be chosen to be solve variables?

3. An equation of the form $ax + by + cz = d$ is the equation of a plane, and the solution of a system of three equations in three unknowns is the common intersection of three planes. Maple can be made to produce a nice picture of this intersection using the max function. (For example, entering `max(3,7,2,4)` will return the largest of the numbers in the list – in this case 7.)

Solve the following system of equations.

$$
\begin{aligned}
x + y + z &= 3 \\
x + y - z &= 1 \\
x - y + z &= 2
\end{aligned}
$$

To get Maple to plot the three planes we need to solve each equation for z.

$$
\begin{aligned}
z &= 3 - x - y \\
z &= x + y - 1 \\
z &= 2 - x + y
\end{aligned}
$$

To see the uppermost portion of the planes, define `z := max(3-x-y, x+y-1, 2-x+y)` and then `plot3d(z, x= -10..10, y= -10..10)`. To see the complete set of planes, plot the set {3-x-y, x+y-1, 2-x+y}. You will see graphs similar to those in Figure 1.5. (Consult Section 2.1.215 of the Maple V Library Reference Manual for further information on `plot3d`.)

From a geometric point of view, what can you say about the solutions of systems
of three equations in three unknowns in case the solution is not unique?

```
> plot3d(z, x=-10..10, y=-10..10, axes=boxed, orientation=[40,70]);
```

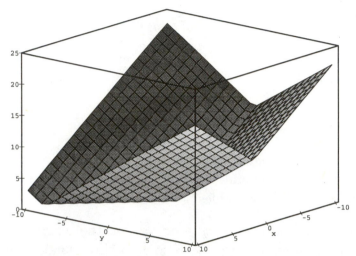

```
> plot3d({2-x+y, x+y-1, 3-x-y}, x=-10..10, y=-10..10,
    axes=boxed, orientation=[40,70]);
```

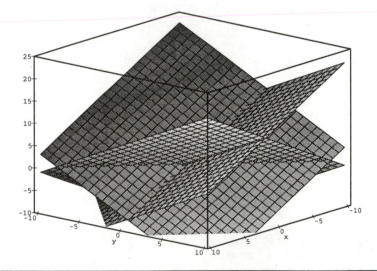

Figure 1.5: The intersection of three planes

1.5   Producing a Dietary Supplement

LABORATORY EXERCISE 1–1

Name _____   **Due Date** _____

The problem: You wish to produce 1000 kilograms of a product containing 0.25% folic acid, 0.25% vitamin B-6, 12% calcium, and 2% iron. This product is to be pressed into tablets and sold as a dietary supplement. The tablets are to be made from some (or all) of six preparations that can be obtained from your distributor. The pertinent information is displayed in the following table.

	%folic acid	%vitamin B-6	%calcium	%iron	cost($ per kilogram)
A	0.1	0.1	18	3	4.20
B	0.3	0	9	3	3.70
C	0.2	0.4	14	1	2.50
D	0.1	0.5	19	1.2	5.60
E	0.4	0.2	10	2	7.20
F	0.2	0.2	11	3	4.50

1. Write a system of equations whose solution provides the appropriate amount of each ingredient.

continued on next page

2. Solve the system of equations.

3. Express the cost in terms of a single variable.

4. What value of the variable used in 3 above gives the minimum cost? Explain carefully how you arrived at this value.

5. How much of each ingredient should you use to obtain a product of minimum cost?

6. If you press the product into 4-gram capsules and sell it in bottles of 250 for $6.00 per bottle, what is your profit?

Chapter 2
AUGMENTED MATRICES AND ELEMENTARY ROW OPERATIONS

LINEAR ALGEBRA CONCEPTS

- **Augmented matrix**
- **Elementary row operation**
- **Echelon form**
- **Reduced echelon form**
- **Gauss-Jordan Elimination**
- **Transpose**
- **Rank (optional)**

2.1 Introduction

A matrix is said to be in *echelon form* if

1. the first nonzero entry in each nonzero row is "1,"
2. all zero rows are below all nonzero rows, and
3. the leading "1" in a nonzero row occurs to the right of the leading "1" in any previous row.

If, in addition, each column with a leading "1" has zeros everywhere else, then the matrix is said to be in *reduced echelon form* (or simply *row-reduced*). Maple's function `rref` returns the reduced echelon form of a matrix.

For the following problems, we will need to use Maple's `linalg` package; make it available by entering `with(linalg)` at the beginning of your Maple session. End the statement with a colon, Maple's "silent terminator," instead of the usual semicolon, to avoid printing the names of all functions defined when the package is loaded. Enter `?linalg` to see a list of all the functions in the package.

2.2   Solved Problems

<u>Solved Problem 1</u>: Display the row operations required to put the matrix below into echelon form.

$$\begin{bmatrix} 0 & 3 & 5 \\ 2 & 4 & 0 \\ 4 & 8 & 1 \end{bmatrix}$$

<u>Solution</u>: Each of the three elementary row operations is included in the `linalg` package that comes with Maple.

First, load the `linalg` package by entering `with(linalg)` followed by a colon. The warnings Maple gives in Figure 2.1 are about functions that have been redefined by `linalg`.

```
                Row Operations from the linalg Package

    mulrow (A, r, s) multiplies row r of matrix A by the number s.
    addrow (A, i, j, s) adds s times row i of matrix A to row j.
    swaprow (A, i, j) swaps rows i and j of matrix A.
```

To enter a matrix into Maple element by element, first set the dimensions with `matrix(r,c)` where r and c are the number of rows and columns, respectively. Now enter `entermatrix('')`. Recall that `''` indicates the previous result. Maple will now prompt you to enter the elements of the matrix; note that each number must be followed by a semicolon. To use this technique with a named matrix, execute, for example, `A := matrix(2,3)` and then `entermatrix(A)`. For details, see Section A.5 in Appendix A.

In line 2 of Figure 2.1, we have entered the matrix in a single statement. Again, for details, see Section A.5 in Appendix A.

To swap rows 1 and 2, enter `swaprow('', 1, 2)`. This is done in line 3 of Figure 2.1. (Alternatively, if the matrix is named, you may refer to the matrix by its name: assuming it is `A`, you could enter `swaprow(A, 1, 2)`.)

The next step is to multiply row 1 by $\frac{1}{2}$ using `mulrow('', 1, 1/2)`. To subtract 4 times row 1 from row 3, enter `addrow('', 3, 1, -4)`. An echelon form appears in line 5 of Figure 2.1. You should be aware that the echelon form of a matrix is not unique. Thus, if you perform a different sequence of row operations, you may arrive at a correct answer that is different from that in Figure 2.1.

```
> with(linalg):
Warning:   new definition for      norm
Warning:   new definition for      trace
> matrix(3, 3, [0,1,3, 2,4,0, 4,8,1]);
```

$$\begin{bmatrix} 0 & 1 & 3 \\ 2 & 4 & 0 \\ 4 & 8 & 1 \end{bmatrix}$$

```
> swaprow(", 1, 2);
```

$$\begin{bmatrix} 2 & 4 & 0 \\ 0 & 1 & 3 \\ 4 & 8 & 1 \end{bmatrix}$$

```
> mulrow(", 1, 1/2);
```

$$\begin{bmatrix} 1 & 2 & 0 \\ 0 & 1 & 3 \\ 4 & 8 & 1 \end{bmatrix}$$

```
> addrow(", 1, 3, -4);
```

$$\begin{bmatrix} 1 & 2 & 0 \\ 0 & 1 & 3 \\ 0 & 0 & 1 \end{bmatrix}$$

Figure 2.1: Step by step row-reduction of a matrix

A Maple Suggestion: Some users, like us, may find it convenient to rename some of these functions to shorten their names, and even change their arguments to make them more consistent with the way elementary row operations are presented in most texts. For example, if you define `scale := (i, s) -> mulrow(A, i, s)`, then `scale(i, s)` will multiply row i of the matrix A by s. (For more information on defining functions, see Appendix A.) *To keep this book consistent with the Maple manuals, we will use the standard definitions.*

Solved Problem 2: Determine if it is possible by elementary row operations to transform the matrix $\begin{bmatrix} 1 & 2 & 3 \\ 4 & 5 & 6 \\ 7 & 8 & 9 \end{bmatrix}$ into the matrix $\begin{bmatrix} 2 & 5 & 2 \\ 1 & 4 & 3 \\ 3 & 2 & 1 \end{bmatrix}$.

Solution: Enter the first matrix as done in Solved Problem 1 above. Next, row-reduce with `rref('')`. The reduced echelon form is line 2 of Figure 2.2. Repeat this procedure to get the reduced echelon form of the second matrix (line 4). The two are obviously different. Since the *reduced* echelon form is unique, unlike the row echelon form, we conclude that it is not possible to obtain the second matrix from the first using elementary row operations.

Maple hint: If you intend to do more than one operation on a given matrix, it is helpful to give it a name. For example, begin line 1 with `A := matrix`, Maple will know that the name A refers to the matrix defined in line 1. You can now get the row-reduced form of A using `rref(A)`.

```
> matrix(3, 3, [1,2,3, 4,5,6, 7,8,9]);
```

$$\begin{bmatrix} 1 & 2 & 3 \\ 4 & 5 & 6 \\ 7 & 8 & 9 \end{bmatrix}$$

```
> rref(");
```

$$\begin{bmatrix} 1 & 0 & -1 \\ 0 & 1 & 2 \\ 0 & 0 & 0 \end{bmatrix}$$

```
> matrix(3, 3, [2,5,2, 1,4,3, 3,2,1]);
```

$$\begin{bmatrix} 2 & 5 & 2 \\ 1 & 4 & 3 \\ 3 & 2 & 1 \end{bmatrix}$$

```
> rref(");
```

$$\begin{bmatrix} 1 & 0 & 0 \\ 0 & 1 & 0 \\ 0 & 0 & 1 \end{bmatrix}$$

Figure 2.2: Matrices with different reduced echelon forms

Solved Problem 3: Use Gauss-Jordan Elimination to solve the following two systems.

$$\begin{array}{rcl} 2x + 3y - 7z + w &=& 8 \\ 4x - 2y + z + 2w &=& 4 \\ 5x + y - 4z + 3w &=& 6 \end{array} \qquad \begin{array}{rcl} 2x + 3y - 7z + w &=& 8 \\ 4x - 2y + z + 2w &=& 4 \\ 5x + y - 4z + 3w &=& 5 \end{array}$$

<u>Solution</u>: The augmented matrix of the first system is

$$\left[\begin{array}{ccc|c} 2 & 3 & -7 & 1 & 8 \\ 4 & -2 & 1 & 2 & 4 \\ 5 & 1 & -4 & 3 & 6 \end{array}\right]$$

To enter this matrix, issue the commands just as in Solved Problem 1. (For alternative methods of entering matrices, see Section A.5 in Appendix A.) When you have finished, the matrix will appear as seen in line 1 of Figure 2.3.

```
> matrix(3, 5, [2,3,-7,1,8, 4,-2,1,2,4, 5,1,-4,3,6]);
```

$$\left[\begin{array}{ccccc} 2 & 3 & -7 & 1 & 8 \\ 4 & -2 & 1 & 2 & 4 \\ 5 & 1 & -4 & 3 & 6 \end{array}\right]$$

```
> rref(");
```

$$\left[\begin{array}{ccccc} 1 & 0 & 0 & \frac{16}{21} & \frac{-10}{21} \\ 0 & 1 & 0 & \frac{5}{7} & \frac{-32}{7} \\ 0 & 0 & 1 & \frac{8}{21} & \frac{-68}{21} \end{array}\right]$$

```
> matrix(3, 5, [2,3,-7,1,8, 4,-2,1,2,4, 6,1,-6,3,5]);
```

$$\left[\begin{array}{ccccc} 2 & 3 & -7 & 1 & 8 \\ 4 & -2 & 1 & 2 & 4 \\ 6 & 1 & -6 & 3 & 5 \end{array}\right]$$

```
> rref(");
```

$$\left[\begin{array}{ccccc} 1 & 0 & \frac{-11}{16} & \frac{1}{2} & 0 \\ 0 & 1 & \frac{-15}{8} & 0 & 0 \\ 0 & 0 & 0 & 0 & 1 \end{array}\right]$$

Figure 2.3: $\boxed{\text{Solving by Gauss-Jordan Elimination}}$

Now row-reduce the matrix. The solution of the system of equations can be read from line 2 of Figure 2.3 by solving for the leading variables.

$$x = -\frac{10}{21} - \frac{16}{21}w$$
$$y = -\frac{32}{7} - \frac{5}{7}w$$
$$z = -\frac{68}{21} - \frac{8}{21}w$$
$$w \quad \text{is arbitrary}$$

To solve the second system of equations, follow the instructions above to enter the augmented matrix

$$\begin{bmatrix} 2 & 3 & -7 & 1 & 8 \\ 4 & -2 & 1 & 2 & 4 \\ 5 & 1 & -4 & 3 & 5 \end{bmatrix}$$

Now row-reduce this matrix. From the last row of the matrix in line 4 of Figure 2.3, we conclude that this system of equations has no solution.

You may wish to compare your work here with what happens if you solve these systems using the methods of Chapter 1.

Solved Problem 4: Find values of p, q, and r so that the following system of equations has no solution.

$$2x + y + 3z + 5w = p$$
$$3x + 2y - z + 4w = q$$
$$3x + y + 10z + 11w = r$$

Solution: Enter the 3 by 5 augmented matrix of this system as described in the solved problems above. Trying to use the reduced row echelon form via **rref** leads to a problem. Line 2 of Figure 2.4 shows the reduced form—the variables have been eliminated. (Why?)

Instead, row-reduce the matrix using **gausselim** (an abreviation/contraction of Gaussian elimination). From the last row of the matrix in line 3 of Figure 2.4, we see that the system of equations has a solution if and only if $r - 3p + q = 0$. Thus, any choice of p, q, and r that does not satisfy this equation will work. One correct solution is $p = q = r = 1$.

```
> A := matrix(3, 5, [2,1,3,5,p, 3,2,-1,4,q, 3,1,10,11,r]);
```

$$A := \begin{bmatrix} 2 & 1 & 3 & 5 & p \\ 3 & 2 & -1 & 4 & q \\ 3 & 1 & 10 & 11 & r \end{bmatrix}$$

```
> rref(A);
```

$$\begin{bmatrix} 1 & 0 & 7 & 6 & 0 \\ 0 & 1 & -11 & -7 & 0 \\ 0 & 0 & 0 & 0 & 1 \end{bmatrix}$$

```
> gausselim(A);
```

$$\begin{bmatrix} 2 & 1 & 3 & 5 & p \\ 0 & \frac{1}{2} & \frac{-11}{2} & \frac{-7}{2} & q - \frac{3}{2}p \\ 0 & 0 & 0 & 0 & r - 3p + q \end{bmatrix}$$

Figure 2.4: Making a system of equations inconsistent

2.3   Exercises

1. Use Gauss-Jordan Elimination to solve the following systems.

   (a)
   $$2x - 3y + 4z - 5w = 6$$
   $$3x + 4y - 2z + w = 4$$
   $$5x - 2y + z + 3w = 8$$

   (b)
   $$3a + 4b - 7c + 6d - e = 1$$
   $$4a + b - 2c + 4d - 3e = 4$$
   $$2a + 3b + c + 4d + 7e = 5$$

   (c)
   $$5x + 4y - z + w = 8$$
   $$2x - y + 2z - w = 3$$
   $$7x + 3y + z = 2$$

2. Find values of p, q, r, and s so that the following system of equations has no solution.

   $$3a + 4b + 5c - d + e = p$$
   $$4a + 2b - c + 7d - e = q$$
   $$2a - 3b + 4c + d + 3e = r$$
   $$5a + 9b + 5d - 3e = s$$

3. Each of the three systems of equations below has a solution. Find a, b, and c.

   $$2x + 3y + 4z = a \qquad 3x - 2y + 7z = a \qquad 5x + 4y - 9z = a$$
   $$4x + 2y - z = b \qquad 2x - 4y + 6z = b \qquad 3x + 6y + z = b$$
   $$6x + 5y + 3z = c \qquad 4x + 8z = c \qquad 2x - 2y - 10z = c$$

4. Solve the following systems of equations.

   (a)
   $$2x^3 + 5y^3 - 7z^3 + w^3 = 5$$
   $$4x^3 - 3y^3 + z^3 - 2w^3 = 2$$
   $$3x^3 + 3y^3 + 5z^3 - w^3 = 1$$
   $$6x^3 - 5y^3 + 4z^3 + 9w^3 = 7$$

   (b)
   $$(a + b + c)^2 + 5(a + b + c) - 2 = 0$$
   $$3a - 2b + 7c = 0$$
   $$5a + 4b + 2c = 9$$

   (Hint: Maple's direct solution of this system is difficult to understand. Consider solving $x^2 + 5x - 2 = 0$ first.)

5. Determine if $\begin{bmatrix} 2 & 3 & 4 & 5 \\ 8 & 2 & 5 & 1 \\ 7 & 7 & 5 & 3 \end{bmatrix}$ can be tranformed into $\begin{bmatrix} 3 & 6 & 9 & 2 \\ 4 & 5 & 0 & 1 \\ 3 & 1 & 5 & 8 \end{bmatrix}$ by elementary row operations.

6. Find a solution of the following system of equations such that x, y, z and w are integers.

$$
\begin{aligned}
2x + y - z + w &= 5 \\
3x - 2y + z - w &= 0 \\
2x + 2y - 2z + w &= 3
\end{aligned}
$$

7. Let $A = \begin{bmatrix} 1 & 2 & 3 & 2 \\ 4 & 5 & 6 & 11 \\ 7 & 8 & 9 & 10 \end{bmatrix}$ and $B = \begin{bmatrix} 1 & 2 & 3 & 1 \\ 4 & 5 & 6 & 11 \\ 7 & 8 & 9 & 10 \end{bmatrix}$.

(a) Find the transposes of the row-reduced forms of A and B. (Maple's syntax for the transpose of A is transpose(A).)

(b) Find the row-reduced forms of the transposes of A and B.

(c) What does this say about the proposition that the reduced form of the transpose is the transpose of the reduced form?

2.4   Exploration and Discovery

1. For this exercise we will define the *rank* of a matrix M to be the number of nonzero rows in the reduced echelon form of M. It is easy to calculate using Maple — just `rref` and count the number of nonzero rows. For example, if

$$M = \begin{bmatrix} 1 & 2 & 3 \\ 4 & 5 & 6 \\ 7 & 8 & 9 \end{bmatrix}, \text{ then the reduced echelon form of } M \text{ is } \begin{bmatrix} 1 & 0 & -1 \\ 0 & 1 & 2 \\ 0 & 0 & 0 \end{bmatrix}, \text{ so the}$$

rank of M is 2. (Other equivalent definitions of "rank" are discussed further in Chapter 8.)

   (a) Do Exercise 7.

   (b) Let A and B be as in Exercise 7. Calculate the ranks of A and B.

   (c) Calculate the ranks of the transposes of A, B, and M above.

   (d) Experiment with other matrices of your choice, including a 3 by 5 and a 4 by 2. Discuss your observations and conclusions. (See `randmatrix` in Appendix B for instructions on how to generate random examples.)

   (e) Suppose that A is a matrix whose rank is equal to the number of rows of A. What can you say about the consistency of a system of equations whose augmented matrix is $[A \mid B]$?

2. **Maximizing a linear function over a convex polygon**. In many applications, it is important to maximize a linear function subject to constraints introduced by a system of linear inequalities. Such problems are referred to as "linear programming" problems. For example, let's maximize the function $C(x, y) = 2x + 3y - 1$ subject to the following system of inequalities:

$$\begin{aligned} x &\geq 0 \\ y &\geq 0 \\ x + y &\leq 1 \end{aligned}$$

Each inequality describes a half-plane (see Figure 2.5). For example, the inequality $x + y \leq 1$ describes the half-plane below the line $x + y = 1$. Linear inequalities taken together describe an intersection of half-planes that is often a convex polygon. The inequalities above, for example, describe the triangle in the plane with vertices $(0, 0)$, $(0, 1)$, and $(1, 0)$. **It is known that the**

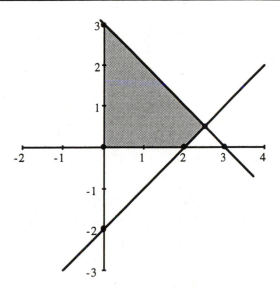

Figure 2.5: A polygon defined by linear inequalities

maximum of the linear function $C(x, y)$ **must occur at a vertex of the polygon perhaps including an edge as well.** Since $C(0,0) = -1$, $C(1,0)=1$ and $C(0,1) = 2$, we conclude that, subject to the constraints above, C achieves its maximum, 2, at the point $(0,1)$.

In general, the difficulty in solving a problem like the one above is in finding the vertices of the polygon. Maple can help. Let's maximize $C(x, y) = 2x + 3y - 1$ above subject to the following constraints.

$$
\begin{aligned}
x &\geq 0 \\
y &\geq 0 \\
x + y &\leq 3 \\
x - y &\leq 2
\end{aligned}
$$

The vertices of the convex region are among the intersections of the lines $x = 0$, $y = 0$, $x + y = 3$, and $x - y = 2$. Thus, we need to solve the following six systems of equations.

$$
\begin{array}{ccc}
x = 0 & x = 0 & x = 0 \\
y = 0 & x + y = 0 & x - y = 2
\end{array}
$$

$$
\begin{array}{ccc}
y = 0 & y = 0 & x + y = 3 \\
x + y = 3 & x - y = 2 & x - y = 2
\end{array}
$$

Enter the 4 by 3 augmented matrix of the system of inequalities
$\begin{bmatrix} 1 & 0 & 0 \\ 0 & 1 & 0 \\ 1 & 1 & 3 \\ 1 & -1 & 2 \end{bmatrix}$

as matrix A. Notice that the augmented matrix of any of the six systems of two equations in two unknowns above can be obtained by selecting two appropriate rows from A. For example, the augmented matrix of

$$
\begin{array}{rcl}
x & = & 0 \\
x + y & = & 3
\end{array}
$$

consists of rows 1 and 3 of A. We need an efficient way to get the solutions of the six systems of equations above. The following function will help. Define the function f by

```
f := (i,j) -> rref(stack(row(A,i), row(A,j)))
```

The function $f(i,j)$ returns the reduced form of the matrix consisting of rows i and j of A. To obtain the solutions of the six systems of equations above, enter the six expressions: f(1,2), f(1,3), f(1,4), f(2,3), f(2,4), and f(3,4). Maple will return reduced augmented matrices from which the following solutions can be read:

$$(0,0),(0,3),(0,-2),(3,0),(2,0),\left(\tfrac{5}{2},\tfrac{1}{2}\right)$$

Not all of these points are vertices of the polygon. (See Figure 2.5.) Only the points $(0,0)$, $(0,3)$, $(2,0)$, and $(\tfrac{5}{2},\tfrac{1}{2})$ satisfy *all four* of the inequalities above. (We find this out by simply checking each of the points in the four inequalities to see if they work.) Finally, $C(0,0) = -1$, $C(0,3) = 8$, $C(2,0) = 3$, and $C(\tfrac{5}{2},\tfrac{1}{2}) = \tfrac{11}{2}$. We conclude that C reaches its maximum value, 8, at the point $(0,3)$.

Find the maximum value of the function $C(x, y)$ above subject to the following inequalities.

$$
\begin{aligned}
x &\geq 0 \\
y &\geq 0 \\
-x + y &\leq 3 \\
x + y &\leq 5 \\
-2x + y &\geq -4
\end{aligned}
$$

2.5 Row Operations

LABORATORY EXERCISE 2–1

Name _____   **Due Date** _____

1. Display the row operations required to put $\begin{bmatrix} 0 & -1 & 0 & 4 & 3 \\ 1 & 5 & -2 & 6 & 5 \\ 8 & 1 & 4 & 7 & 7 \end{bmatrix}$ into echelon

   form, as we did in Solved Problem 1.

2. Determine if the matrix from 1 can be transformed into $\begin{bmatrix} 32 & 3 & 16 & 32 & 31 \\ 97 & 16 & 46 & 94 & 92 \\ 72 & 7 & 36 & 71 & 69 \end{bmatrix}$

   by elementary row operations. Explain.

3. Determine if the matrix from 1 can be transformed into $\begin{bmatrix} 97 & 16 & 46 & 94 & 92 \\ 32 & 3 & 16 & 32 & 31 \\ 72 & 17 & 36 & 71 & 69 \end{bmatrix}$

   by elementary row operations. Explain.

2.6 Maximizing a Linear Function over a Polygon

LABORATORY EXERCISE 2–2

Name _____   **Due Date** _____

The problem: Find the maximum of the function $C(x,y,z) = 3x + y + 3z$ over the polygon determined by the following inequalities.

$$
\begin{aligned}
x &\geq 0 \\
y &\geq 0 \\
z &\geq 0 \\
2x + y + z &\leq 2 \\
x + 2y + 3z &\leq 5 \\
2x + 2y + z &\leq 6
\end{aligned}
$$

Discussion: This problem is similar to Exploration and Discovery Problem 2 except that it is set in one higher dimension. Here the convex polygon is the intersection of half-spaces defined by the inequalities, and the vertices are intersections of triples of planes. In Exploration and Discovery Problem 2, we defined a function f(i,j) to select pairs of rows of the augmented matrix and row-reduce them. Here we need to solve systems of three equations in three unknowns. Thus, we need to select triples of rows from

$$
A = \begin{bmatrix}
1 & 0 & 0 & 0 \\
0 & 1 & 0 & 0 \\
0 & 0 & 1 & 0 \\
2 & 1 & 1 & 2 \\
1 & 2 & 3 & 5 \\
2 & 2 & 1 & 6
\end{bmatrix}.
$$

For that, another variable must be added. Define the new function as follows:

```
f := (i,j,k) -> rref(stack(row(A,i), row(A,j), row(A,k)))
```

continued on next page

1. Find the intersection of each triple of planes defined by the following equations.

$$
\begin{aligned}
x &= 0 \\
y &= 0 \\
z &= 0 \\
2x + y + z &= 2 \\
x + 2y + 3z &= 5 \\
2x + 3y + z &= 6
\end{aligned}
$$

continued on next page

2. Which of the points found in 1 above are vertices of the polygon defined by the given inequalities?

3. Find the maximum of $C(x, y, z)$ subject to the constraints above.

4. You are to maximize a linear function of 100 variables subject to 500 inequalities each involving 100 variables. Thus, in order to find the vertices of the polygon, you must solve many systems of 100 equations in 100 unknowns. The total number of such systems is given by $\frac{500!}{100! \ 400!}$, which in Maple syntax is 500!/(100!*400!).

Suppose that your computer can solve a system of 100 equations in 100 unknowns in 5 seconds. If you begin your calculations today, when will this project be completed? (Give the month, day, and year.)

Chapter 3
THE ALGEBRA OF MATRICES

LINEAR ALGEBRA CONCEPTS

- **Matrix addition**
- **Scalar multiplication**
- **Matrix multiplication**

3.1 Introduction

Matrices are far more important in mathematics than simply serving as an aid in solving systems of linear equations. They are studied as objects in their own right with a natural algebraic structure similar in many ways to that of real numbers, but there are also striking differences that we will explore.

Important Maple Notes: We usually don't distinguish between the 1 by 3 matrix [1 2 3] and the vector $(1, 2, 3)$, but Maple does. Also, matrix multiplication is different than ordinary multiplication, so Maple uses &* for matrix products. The &* is essential; neither A*B, nor AB is acceptable. You should consult Section A.8 in Appendix A for details before proceeding.

3.2 Solved Problems

Solved Problem 1: Set $A = \begin{bmatrix} 1 & 2 & 3 \\ 2 & 3 & 0 \\ 4 & 8 & 5 \end{bmatrix}$ and $B = \begin{bmatrix} 3 & 8 & -2 \\ 4 & 7 & -1 \\ 0 & 3 & 5 \end{bmatrix}$ and then calculate

$3BA - 4A^3$.

Solution: The first step we take is to load the linear algebra routines by executing `with(linalg)`. Remember that the warnings (line 1, Figure 3.1) just indicate that `linalg` redefines these functions. Now enter the matrix A. It will be convenient to name the matrix, so begin the statement with `A := matrix` as in line 2 of Figure 3.1. The ":=" is Maple's syntax for a *definition* rather than an equation. Maple now knows that the name A refers to the matrix you have just entered, and this definition will remain in effect until you change it. Use the same technique to enter the second

```
> with(linalg):
```
Warning: new definition for   norm
Warning: new definition for   trace
```
> A := matrix(3,3, [1,2,3, 2,3,0, 4,8,5]);
```

$$A = \begin{bmatrix} 1 & 2 & 3 \\ 2 & 3 & 0 \\ 4 & 8 & 5 \end{bmatrix}$$

```
> B := matrix(3,3, [3,8,-2, 4,7,-1, 0,3,5]);
```

$$B = \begin{bmatrix} 3 & 8 & -2 \\ 4 & 7 & -1 \\ 0 & 3 & 5 \end{bmatrix}$$

```
> 3*B &* A - 4*A^3;
```

$$((3B) \ \&^* \ A) - 4A^3 \qquad\qquad 3BA - 4A^3$$

```
> evalm("");
```

$$\begin{bmatrix} -579 & -1054 & -567 \\ -190 & -349 & -195 \\ -1250 & -2221 & -1145 \end{bmatrix}$$

Figure 3.1: $\boxed{\text{Calculating with matrices}}$

matrix and name it B.

Maple uses &* to denote matrix multiplication. Thus, we enter 3*B &* A - 4*A^3. Maple does not automatically evaluate this expression. We must specifically request matrix evaluation with evalm, which is short for evaluate as a matrix. (To do this in one step, we could have entered evalm(3*B &* A - 4*A^3).) The result is line 5 of Figure 3.1.

Solved Problem 2: Let $A = \begin{bmatrix} 1 & 2 & 3 \\ 4 & 5 & 6 \\ 7 & 8 & 9 \end{bmatrix}$ and $f(x) = x^4 - 3x^2 + 5$. Calculate $f(A)$.

Solution: We must first agree on what $f(A)$ means. If we formally replace x in the definition of f by A (e.g., using the substitute command), we obtain $A^4 - 3A^2 + 5$, which requires that we add the number 5 to the matrix $A^4 - 3A^2$; but we can't add a number to a matrix. The standard convention in evaluating polynomials at matrices is to replace the constant term by the constant times the identity matrix. Thus, $f(A)$ is defined to be $A^4 - 3A^2 + 5I$, where I is the 3 by 3 identity matrix. Maple's syntax for the n by n identity matrix is &*(). This is very strange notation. Fortunately, Maple adheres to the standard convention and evaluates A^4 - 3 * A^2 + 5 as if the 5 were $5I$, or, in Maple, as if the 5 were 5* &*().

From this point on, we will assume that you have already loaded the linalg package. Enter the matrix and name it A as seen in Figure 3.2. Next, enter A^4 - 3 * A^2 + 5. Just as before, we must apply evalm to have Maple fully calculate this expression. The answer appears in line 3 of Figure 3.2.

```
> A := matrix(3,3, [1,2,3, 4,5,6, 7,8,9]);
```

$$A = \begin{bmatrix} 1 & 2 & 3 \\ 4 & 5 & 6 \\ 7 & 8 & 9 \end{bmatrix}$$

```
> A^4 - 3 * A^2 + 5;
```

$$A^4 - 3A + 5$$

```
> evalm(");
```

$$\begin{bmatrix} 7475 & 9180 & 10890 \\ 16920 & 20795 & 24660 \\ 26370 & 32400 & 38435 \end{bmatrix}$$

Figure 3.2: A polynomial evaluated at a matrix

Solved Problem 3: For students who have studied calculus. Define the matrices

$$D = \begin{bmatrix} 1 & 0 & 0 \\ 0 & 0 & 0 \\ 0 & 0 & 0.6 \end{bmatrix} \qquad A = \begin{bmatrix} \frac{4}{11} & \frac{17}{11} & -\frac{13}{11} \\ -\frac{14}{11} & \frac{45}{11} & -\frac{26}{11} \\ -\frac{35}{22} & \frac{85}{22} & -\frac{43}{22} \end{bmatrix}$$

Find $\lim_{n \to \infty} D^n$ and $\lim_{n \to \infty} A^n$.

<u>Discussion:</u> An explanation is in order here. Just exactly what does $\lim_{n \to \infty} M^n$ mean when M is a matrix?

Definition: $\lim_{n \to \infty} M^n = P$ if $\lim_{n \to \infty} m_{i,j}(n) = p_{i,j}$ where $m_{i,j}(n)$ is the (i,j)th term of M^n.

In other words, take the limit of each entry of M^n.

Solution: You should be able to convince yourself that $\begin{bmatrix} a & 0 & 0 \\ 0 & b & 0 \\ 0 & 0 & c \end{bmatrix}^n = \begin{bmatrix} a^n & 0 & 0 \\ 0 & b^n & 0 \\ 0 & 0 & c^n \end{bmatrix}$

for any positive integer n; you can verify it with Maple. Therefore,

$$\lim_{n \to \infty} D^n = \lim_{n \to \infty} \begin{bmatrix} 1 & 0 & 0 \\ 0 & 0 & 0 \\ 0 & 0 & 0.6^n \end{bmatrix} = \begin{bmatrix} 1 & 0 & 0 \\ 0 & 0 & 0 \\ 0 & 0 & \lim_{n \to \infty} 0.6^n \end{bmatrix} = \begin{bmatrix} 1 & 0 & 0 \\ 0 & 0 & 0 \\ 0 & 0 & 0 \end{bmatrix}$$

That was easy. Now, what about A? Unlike the simple diagonal case, there's no apparent easy formula for A^n. We can't just take the nth powers of the elements. So it seems we have no way of actually calculating this limit or even knowing if it exists! But we can look at large powers of A in order to develop some intuition regarding this limit. In Chapter 15 we will see that under the right conditions we can calculate such a limit exactly, but for now we can only estimate it. Enter the matrix and name it A. We will look at A^{10}, A^{20}, A^{30}, and A^{40}. If we evaluate A^10, A^20, A^30, and A^40 as matrix expressions, Maple will respond with the *exact* value, which will be difficult to analyze and compare. We can approximate the values with Maple's evalf (short for evaluate as floating point). To approximate A^{20} directly, compose evalf with evalm, as seen in line 3 of Figure 3.3. In the same way, we've approximated A^{30} and A^{40} in lines 4 and 5. The corresponding entries in A^{30} and A^{40} are the same to eight digits of accuracy, so it seems reasonable that these matrices are good approximations to the desired limit.

Solved Problem 4: If $A = \begin{bmatrix} 2 & 3 & 4 \\ 6 & 4 & 2 \\ 3 & 1 & 6 \end{bmatrix}$, find a polynomial $f(x) = x^3 + px^2 + qx + r$ such that $f(A) = 0$.

Solution: Enter and name the matrix A. Next enter the expression A^3 + p*A^2 + q*A + r. Now evaluate the last result as a matrix and Maple will present the matrix in line 3 of Figure 3.4. We want each entry in this matrix to be 0. Thus, we have a system of nine equations in three unknowns to solve. To do this with Maple, we need to list the nine entries of the matrix as the elements of a set. The Maple function convert will do this.

```
> evalm(A^10);
```

$$\begin{bmatrix} \frac{-139}{512} & \frac{1581}{512} & \frac{-1209}{512} \\ \frac{-651}{256} & \frac{1837}{256} & \frac{-1209}{256} \\ \frac{-3255}{1024} & \frac{7905}{1024} & \frac{-5051}{1024} \end{bmatrix}$$

```
> evalf(");
```

$$\begin{bmatrix} -.271484375 & 3.087890625 & -2.361328125 \\ -2.54296875 & 7.17578125 & -4.72265625 \\ -3.178710938 & 7.719726563 & -4.903320313 \end{bmatrix}$$

```
> evalf(evalm(A^20));
```

$$\begin{bmatrix} -.272726059 & 3.090906143 & -2.363634109 \\ -2.545452118 & 7.181812286 & -4.727268219 \\ -3.181815147 & 7.727265358 & -4.909085274 \end{bmatrix}$$

```
> evalf(evalm(A^30));
```

$$\begin{bmatrix} -.2727272715 & 3.090909088 & -2.363636361 \\ -2.545454543 & 7.181818176 & -4.727272723 \\ -3.181818179 & 7.72727272 & -4.909090904 \end{bmatrix}$$

```
> evalf(evalm(A^40));
```

$$\begin{bmatrix} -.2727272727 & 3.090909091 & -2.363636364 \\ -2.545454545 & 7.181818182 & -4.727272727 \\ -3.181818182 & 7.727272727 & -4.909090909 \end{bmatrix}$$

Figure 3.3: Large powers of a matrix

Enter convert('', set) to obtain the set of equations in line 4 of Figure 3.4. Ask Maple to solve, and the solution can be read from line 5 as $f(x) = x^3 - 12x^2 + 12x + 70$. You should check this answer to ensure that $f(A) = 0$. Solved Problem 2 explains how to evaluate a polynomial at a matrix.

Note: There is something quite remarkable about this result. We solved a system of nine equations in three unknowns and got a unique solution! This says that there is one and only one cubic polynomial f (with leading coefficient 1) for which $f(A) = 0$. We'll return to this topic in a later chapter. See the exercises for further development.

```
> A := matrix(3, 3, [2,3,4, 6,4,2, 3,1,6]);
```

$$A = \begin{bmatrix} 2 & 3 & 4 \\ 6 & 4 & 2 \\ 3 & 1 & 6 \end{bmatrix}$$

```
> A^3+p*A^2+q*A+r;
```

$$A^3 + pA^2 + qA + r$$

```
> evalm("");
```

$$\begin{bmatrix} 314 + 34p + 2q + r & 228 + 22p + 3q & 408 + 38p + 4q \\ 432 + 42p + 6q & 314 + 36p + 4q + r & 504 + 44p + 2q \\ 324 + 30p + 3q & 216 + 19p + q & 458 + 50p + 6q + r \end{bmatrix}$$

```
> convert('', set);
```

$\{314 + 34p + 2q + r,\ 228 + 22p + 3q,\ 408 + 38p + 4q,\ 432 + 42p + 6q,\ 314 + 36p$
$+ 4q + r,\ 504 + 44p + 2q,\ 324 + 30p + 3q,\ 216 + 19p + q,\ 458 + 50p + 6q + r\}$

```
> solve("");
```

$$\{q = 12, r = 70, p = -12\}$$

Figure 3.4: $\boxed{\text{A polynomial that annihilates a matrix}}$

3.3 Exercises

1. If $A = \begin{bmatrix} 1 & 2 & 3 & 0 \\ -1 & 1 & 0 & -2 \\ 2 & 3 & 1 & 0 \\ 4 & 2 & 6 & -3 \end{bmatrix}$ and $B = \begin{bmatrix} 3 & -1 & 2 & 4 \\ 0 & 1 & 2 & 3 \\ -4 & 2 & 0 & -1 \\ 2 & 1 & 4 & 3 \end{bmatrix}$, calculate the follow-

   ing. Comment on how the result compares to the case when A and B are replaced by numbers. (See Exploration and Discovery Problem 1.)

   (a) $A^2 B^2 - (AB)^2$

   (b) $ABA - BA^2$

   (c) $A^2 + 2A + I - (A + I)^2$

   (Note: You can choose I to be the identity matrix &*() or 1. A 4 by 4 identity matrix can also be defined with I4 := band([1],4).)

2. Evaluate the following polynomials at A, where A is as in Exercise 1.

   (a) $x^2 - 3$

   (b) $(x^2 - 3)^2$

   (c) $x^4 - 6x^2 + 9$

   (d) Discuss the relation between the results in (b) and (c).

3. For students who have studied calculus. Approximate $\lim_{n \to \infty} A^n$ in each case and discuss your reason for choosing the value of n you decided to use for the approximation.

   (a) $A = \begin{bmatrix} \frac{21}{4} & -\frac{7}{2} & \frac{13}{8} \\ \frac{21}{4} & -\frac{7}{2} & \frac{17}{8} \\ -\frac{9}{2} & 3 & \frac{1}{4} \end{bmatrix}$

   (b) $B = \begin{bmatrix} 23 & -16 & 7 \\ 33 & -23 & \frac{21}{2} \\ 0 & 0 & 1 \end{bmatrix}$

   (c) $C = \begin{bmatrix} \frac{1}{7} & \frac{2}{7} & \frac{3}{7} \\ \frac{4}{7} & \frac{3}{7} & \frac{5}{7} \\ \frac{6}{7} & 0 & \frac{2}{7} \end{bmatrix}$

4. $\boxed{\text{For students who have studied calculus.}}$ Estimate $\lim_{n \to \infty} \left(I + \frac{A}{n} \right)^n$ where A is $\begin{bmatrix} -7 & 46 & -21 \\ -12 & 79 & -36 \\ -24 & 158 & -72 \end{bmatrix}$, and I is the 3 by 3 identity matrix. Discuss your reason for choosing the value of n you decided to use for the approximation. Having had calculus, you should know what $\lim_{n \to \infty} \left(1 + \frac{x}{n} \right)^n$ is if x is a real number. What is it?

5. Let A be the matrix in Solved Problem 4.

   (a) Show there is no quadratic polynomial $g(x) = x^2 + px + q$ such that $g(A) = 0$.

   (b) Find a polynomial of the form $f(x) = x^4 + px^3 + qx^2 + rx + s$ such that $f(A) = 0$.

   (c) Is there a unique answer to (b)? If not, describe them all.

   (d) Carefully examine your answer to (c) and compare it to the cubic polynomial we found in Solved Problem 4. Explain what you notice.

6. Describe all 2 by 2 matrices that commute with $A = \begin{bmatrix} 2 & -1 \\ 3 & 5 \end{bmatrix}$.

   (Hint. Let $B = \begin{bmatrix} p & q \\ r & s \end{bmatrix}$ and consider $AB - BA = 0$.)

7. Describe all 3 by 3 matrices that commute with $\begin{bmatrix} 1 & 2 & -1 \\ 3 & 0 & 1 \\ 5 & -1 & 1 \end{bmatrix}$.

   (Hint. Let $B = \begin{bmatrix} p & q & r \\ s & t & u \\ v & w & x \end{bmatrix}$ and consider $AB - BA = 0$.)

8. Describe all 3 by 3 matrices that commute with $\begin{bmatrix} -7 & 3 & -3 \\ 30 & 2 & 6 \\ 45 & -9 & 17 \end{bmatrix}$.

9. **Markov chains:** Each year a certain number of people move from city A to city B, while others move from city B to city A. Statistically, it is found that the

probability of a citizen of city A moving to city B in a given year is 0.04, while the probability of a citizen of city B moving to city A is 0.03. For purposes of this exercise we assume that no other factors influence the populations. Thus, if a_n represents the number of people in city A in year n and b_n represents the number of people in city B, then in year $n+1$ we expect to find $0.96a_n + 0.03b_n$ people in city A and $0.04a_n + 0.97b_n$ people in city B.

This says that if $A = \begin{bmatrix} 0.96 & 0.03 \\ 0.04 & 0.97 \end{bmatrix}$, then $A \begin{bmatrix} a_n \\ b_n \end{bmatrix} = \begin{bmatrix} a_{n+1} \\ b_{n+1} \end{bmatrix}$.

Processes of this sort in which the state at time $n+1$ depends only on the state at time n are known as *Markov chains*. A is known as the *transition matrix*. (See *Elementary Linear Algebra, Applications Version* by Howard Anton and Chris Rorres for a precise definition as well as a deeper analysis of the topic.)

(a) Show that in general $A^n \begin{bmatrix} a_0 \\ b_0 \end{bmatrix} = \begin{bmatrix} a_n \\ b_n \end{bmatrix}$.

In the next three parts, assume that in 1990 the population of city A was 240,000 and the population of city B was 196,000.

(b) What population figures would you predict for the year 1995?

(c) What do you expect the population to be in the years 2000 and 2010?

(d) What population figures do you expect the cities will reach far into the future?

10. **More on Markov chains:** Probabilities for population movements among three cities are given in the table below.

	From A	From B	From C
Probability of moving to A	0.7	0.1	0.05
Probability of moving to B	0.2	0.8	0.05
Probability of moving to C	0.1	0.1	0.90

The initial population of city A is 375,000. Find the populations of cities B and C if the total population of the three cities remains constant from year to year.

3.4   Exploration and Discovery

1. The following is a list of familiar formulas that are true for real numbers. Decide which are true for matrices by testing each of them using 2 by 2 matrices with variable entries. Provide specific numerical counterexamples for those that are false and try to prove those that are true. (Compare the first three of them to Exercise 1.)

   (a) $AB = BA$

   (b) $A^2 B^2 = (AB)^2$

   (c) $A^2 + 2A + I = (A + I)^2$

   (d) $(A + B)^2 = A^2 + 2AB + B^2$

   (e) $(A + B)(A - B) = A^2 - B^2$

   (f) $A^2 A^3 = A^3 A^2$

   (g) Test other familiar formulas.

2. Let A be an n by k matrix, let Id denote the n by n identity matrix, and let B denote the augmented matrix $[A \mid Id]$. The command gausselim(B) will return an augmented matrix $[C \mid D]$ where C is the reduced echelon form of A. We wish to investigate the nature of the matrix D. For example, if $A =$

   $$\begin{bmatrix} 1 & 2 & 3 \\ 4 & 5 & 6 \\ 7 & 8 & 9 \end{bmatrix}, \text{ then } D = \begin{bmatrix} 0 & -\frac{8}{3} & \frac{5}{3} \\ 0 & \frac{7}{3} & -\frac{4}{3} \\ 1 & -2 & 1 \end{bmatrix}.$$ (To check this result enter the matrix

   $$\begin{bmatrix} 1 & 2 & 3 & 1 & 0 & 0 \\ 4 & 5 & 6 & 0 & 1 & 0 \\ 7 & 8 & 9 & 0 & 0 & 1 \end{bmatrix}$$ and name it B. Next, execute gausselim(B).) Calculate the product DA for this example.

   For several other choices of A (including matrices that are not square) calculate the product DA. Based on your work, formulate a conjecture about the matrix D and the reduced echelon form of A.

3. $\boxed{\text{For students who have studied calculus.}}$ We've discussed what it means to evaluate a polynomial at a matrix, but what does $\cos(A)$ mean if A is a matrix? (You can ask Maple, but the answer given is $\cos(x)$ applied to the elements of A.) Here's an idea: $\cos(x)$ has a Maclaurin series expansion

$$\cos(x) = 1 - \frac{x^2}{2!} + \frac{x^4}{4!} - \frac{x^6}{6!} \pm \cdots$$

whose partial sums are polynomials that *can* be evaluated at a matrix. Perhaps these can provide an approximation to "$\cos(A)$." As an example, we will approximate $\cos(A)$, where A is $\begin{bmatrix} -7 & 46 & -21 \\ -12 & 79 & -36 \\ -24 & 158 & -72 \end{bmatrix}$.

Enter the matrix and name it A as in line 1 of Figure 3.5. Now use Maple's **taylor** command to obtain a Maclaurin series for $\cos(x)$. This series has a term $O(x^7)$ which must be removed; **convert** the series to a polynomial

---

```
> A := matrix(3, 3, [-7,46,-21, -12,79,-36, -24,158,-72]);
```

$$A = \begin{bmatrix} -7 & 46 & -21 \\ -12 & 79 & -36 \\ -24 & 158 & -72 \end{bmatrix}$$

```
> taylor(cos(x), x, 0, 7);
```

$$1 - \frac{1}{2}x^2 + \frac{1}{24}x^4 - \frac{1}{720}x^6 + O\left(x^7\right)$$

```
> convert(", polynom);
```

$$1 - \frac{1}{2}x^2 + \frac{1}{24}x^4 - \frac{1}{720}x^6$$

```
> subs(x=A, ");
```

$$1 - \frac{1}{2}A^2 + \frac{1}{24}A^4 - \frac{1}{720}A^6$$

```
> evalf(evalm("));
```

$$\begin{bmatrix} .5402777778 & 2.758333333 & -1.379166667 \\ 0 & .5402777778 & 0 \\ 0 & -.9194444444 & 1 \end{bmatrix}$$

Figure 3.5: | Approximating the cosine of a matrix |

as seen in line 3 of Figure 3.5. Next, use `subs` to replace x with A as done in line 4 of Figure 3.5. Finally, approximate with `evalf` to obtain line 5.

(a) Repeat the process above using the Maclaurin polynomial of degrees 10 and 20. (<u>Note</u>: Depending on your machine, it may take several seconds for Maple to do this.) How do the three compare?

(b) Repeat the process above to approximate $\sin(A)$ using Maclaurin polynomials of degrees 6, 10, and 20. (Explain why you don't see an x^6 term in the polynomial of degree 6.)

(c) Square the matrices above that you used to approximate $\cos(A)$ and $\sin(A)$ and add them. Does the formula $\sin^2(x) + \cos^2(x) = 1$ (or one like it) appear to hold for matrices? Explain.

(d) Repeat the process above to approximate e^A using Maclaurin polynomials of degrees 6, 10, and 20. Compare your answers to the one you obtained in Exercise 4.

(e) Does the formula $e^{2x} = (e^x)^2$ appear to hold when x is a matrix?

4. (a) Do Exercise 6.

   (b) If A is the matrix in exercise 6, use Maple to find $f(A)$ where $f(x) = px^2 + qx + r$. (Use variable coefficients as shown.) Using your answer to this and part (a) show that every matrix that commutes with A is a quadratic polynomial in A.

   <u>Remark</u>: It's easy to see that if $f(x)$ is a polynomial and A is a matrix then $f(A)$ commutes with A. The remarkable thing is that the converse is true in this case.

   (c) Do Exercise 7.

   (d) Let A be the matrix from Exercise 7. Is it true that every matrix that commutes with A is a quadratic polynomial in A? If it is true, prove it. If not, give a counter example and prove your example is not a quadratic polynomial in A.

5. | For students who have studied calculus. | The Maple command to produce the derivative of the function $f(x)$ is `diff(f(x), x)`. To ask Maple to differentiate each element of a matrix, we use the `map` command. For example, define the

matrix $A := \begin{bmatrix} x & x^2 \\ x^3 & \sin(x) \end{bmatrix}$. Now differentiate the matrix with respect to x by entering `map(diff, A, x)`. Maple responds with $\begin{bmatrix} 1 & 2x \\ 3x^2 & \cos(x) \end{bmatrix}$. To shorten the typing, you can make you own matrix differentiating function `diffm` with

```
diffm := (A,x) -> map(diff, A, x)
```

Make up two 3 by 3 matrices whose entries are functions of x and test the formula $\frac{d}{dx}AB = (\frac{d}{dx}A)B + A(\frac{d}{dx}B)$. Does this usual product rule for derivatives appear to hold for matrices? Can you prove it? Try similar tests for other differentiation formulas such as the chain rule. Discuss your findings.

3.5   Polynomials and Matrices

LABORATORY EXERCISE 3–1

Name _____   **Due Date** _____

The problem: In algebra you are given problems such as this: Show that $f(1) = 0$ if $f(x) = x^2 + 3x - 4$. As we have seen, we can ask about $f(A)$ when A is a matrix. In Solved Problem 4 and Exercise 5, we examined some questions about finding polynomials f such that $f(A) = 0$. (In this context, 0 means the zero *matrix*, of course.) In this problem we want to examine this idea. In each part of this lab exercise, be sure to discuss how you arrived at your answer. You may also want to attach printouts of Maple screens to refer to in your explanations. Let

$$A = \begin{bmatrix} 1 & 2 & 3 & 0 \\ -1 & 1 & 0 & -2 \\ 2 & 3 & 1 & 0 \\ 4 & 2 & 6 & -3 \end{bmatrix}$$

1. Is there a quadratic polynomial $h(x) = x^2 + px + q$ such that $h(A) = 0$?

2. Is there a cubic polynomial $g(x) = x^3 + px^2 + qx + r$ such that $g(A) = 0$?

continued on next page

3. Find a polynomial $f(x) = x^4 + px^3 + qx^2 + rx + s$ such that $f(A) = 0$.

4. In the cases above where such a polynomial exists, is it unique? If not, describe all polynomials f for which $f(A) = 0$.

5. Based on Solved Problem 4 and your work here, venture a guess about how the size of a square matrix M and the degree of a polynomial $p(x)$ are related when $p(M) = 0$.

6. Let $B = \begin{bmatrix} 2 & 0 \\ 0 & 2 \end{bmatrix}$ and $k(x) = x - 2$. Find $k(B)$. Does this suggest a modification of your conjecture?

3.6   A Genetically Transmitted Plant Disease

LABORATORY EXERCISE 3–2

Name _____   **Due Date** _____

The problem: A field of plants is infected with a genetically transmitted disease. After n years there are x_n diseased plants, y_n plants that are carriers of the disease but do not show any symptoms of the disease, and z_n disease-free plants. (Initially there are x_0 diseased plants, y_0 carriers, and z_0 healthy plants.) We assume that each plant produces a single offspring each year. Half the offspring of the diseased plants die, and the remaining half are diseased. The disease-free plants produce disease-free offspring. The carriers produce $\frac{1}{3}$ diseased plants, $\frac{1}{3}$ carriers, and $\frac{1}{3}$ healthy plants. The goal of this exercise is to trace the course of the disease.

1. Find a formula for each of x_{n+1}, y_{n+1}, and z_{n+1} in terms of x_n, y_n, and z_n.

2. Find a matrix A such that $A \begin{bmatrix} x_n \\ y_n \\ z_n \end{bmatrix} = \begin{bmatrix} x_{n+1} \\ y_{n+1} \\ z_{n+1} \end{bmatrix}$.

continued on next page

3. Use a matrix equation to show how to get $\begin{bmatrix} x_n \\ y_n \\ z_n \end{bmatrix}$ from $\begin{bmatrix} x_0 \\ y_0 \\ z_0 \end{bmatrix}$.

4. Describe the field after 10 years. That is, how many plants are diseased, how many are carriers, and how many are disease-free?

5. Describe the field after 20 years.

6. From the data in 4 and 5 above, what will the field *eventually* look like? That is, after many years, what can you say about the numbers of diseased plants, carriers, and healthy plants?

Chapter 4
INVERSES OF MATRICES

LINEAR ALGEBRA CONCEPTS

- **Matrix inversion**

4.1 Introduction

If a is a non-zero number, its multiplicative inverse, or reciprocal, is the number b such that $ab = 1$. Similarly, if A is a square matrix, its inverse, or reciprocal, is a matrix B such that $AB = BA = I$, if such a matrix exists. As we will see, this idea is important for matrices just as it is for numbers.

4.2 Solved Problems

Solved Problem 1: Find the inverses of $A = \begin{bmatrix} 1 & 2 & 4 \\ 3 & 5 & 9 \\ 2 & 6 & 7 \end{bmatrix}$ and $B = \begin{bmatrix} 1 & 2 & 3 \\ 4 & 5 & 6 \\ 7 & 8 & 9 \end{bmatrix}$.

Solution: First, load the Linear Algebra package: `with(linalg)`. Now enter the two matrices naming them A and B as seen in Figure 4.1. Evaluate `A^(-1)` as a matrix. The inverse is shown in line 3 of Figure 4.1. If we do the same for B, Maple returns an error message (line 4) indicating that no inverse exists. A matrix having no inverse is called *singular*.

Alternative Solution: We can use Maple to emulate the familiar calculation of inverses by row-reducing the augmented matrices $[A \mid I]$ and $[B \mid I]$. Combining the `rref` and `augment` commands allows us to row-reduce the augmented matrices in a single step. It will be helpful to define a 3 by 3 identity matrix with `Id3 := band([1],3)`; see line 1 of Figure 4.2.

To get the inverse of A, define the augmented matrix `AI := augment(A,Id3)`, then row-reduce AI with `rref(AI)`. From line 3 of Figure 4.2, we see that A reduces to the identity matrix. We conclude that A has an inverse, and it appears as the last three columns in line 3. If we perform the same procedure for B, we see from line 5

```
> A := matrix(3,3, [1,2,4, 3,5,9, 2,6,7]);
```

$$A := \begin{bmatrix} 1 & 2 & 4 \\ 3 & 5 & 9 \\ 2 & 6 & 7 \end{bmatrix}$$

```
> B := matrix(3,3, [1,2,3 4,5,6, 7,8,9]);
```

$$B := \begin{bmatrix} 1 & 2 & 3 \\ 4 & 5 & 6 \\ 7 & 8 & 9 \end{bmatrix}$$

```
> evalm(A^(-1));
```

$$\begin{bmatrix} -\frac{19}{7} & \frac{10}{7} & -\frac{2}{7} \\ -\frac{3}{7} & -\frac{1}{7} & \frac{3}{7} \\ \frac{8}{7} & -\frac{2}{7} & -\frac{1}{7} \end{bmatrix}$$

```
> evalm(B^(-1));
Error, (in linalg[inverse]) singular matrix
```

Figure 4.1: Direct method for finding inverses

of Figure 4.2 that the reduced form of B is not the identity matrix. Hence B has no inverse.

Solved Problem 2: Let A and B denote the matrices in Solved Problem 1. Find a 3 by 3 matrix C that satisfies the following equation.

$$A^2(4C - 3B)A^{-1} = BA - AB$$

Solution: The first step is to solve the equation by hand. The steps are shown below. Notice that when we multiply the equation by a matrix we must use the same order of multiplication on each side.

$$\begin{aligned} A^2(4C - 3B)A^{-1} &= BA - AB \\ A^2(4C - 3B) &= (BA - AB)A \\ 4C - 3B &= A^{-2}(BA - AB)A \\ 4C &= A^{-2}(BA - AB)A + 3B \\ C &= \tfrac{1}{4}(A^{-2}(BA - AB)A + 3B) \end{aligned}$$

```
> Id3 := band([1], 3);
```

$$Id3 := \begin{bmatrix} 1 & 0 & 0 \\ 0 & 1 & 0 \\ 0 & 0 & 1 \end{bmatrix}$$

```
> AI := augment(A, Id3);
```

$$AI := \begin{bmatrix} 1 & 2 & 4 & 1 & 0 & 0 \\ 3 & 5 & 9 & 0 & 1 & 0 \\ 2 & 6 & 7 & 0 & 0 & 1 \end{bmatrix}$$

```
> rref(AI);
```

$$\begin{bmatrix} 1 & 0 & 0 & -\frac{19}{7} & \frac{10}{7} & -\frac{2}{7} \\ 0 & 1 & 0 & -\frac{3}{7} & -\frac{1}{7} & \frac{3}{7} \\ 0 & 0 & 1 & \frac{8}{7} & -\frac{2}{7} & -\frac{1}{7} \end{bmatrix}$$

```
> BI := augment(B, Id3);
```

$$BI := \begin{bmatrix} 1 & 2 & 3 & 1 & 0 & 0 \\ 4 & 5 & 6 & 0 & 1 & 0 \\ 7 & 8 & 9 & 0 & 0 & 1 \end{bmatrix}$$

```
> rref(BI);
```

$$\begin{bmatrix} 1 & 0 & -1 & 0 & -\frac{8}{3} & \frac{5}{3} \\ 0 & 1 & 2 & 0 & \frac{7}{3} & -\frac{4}{3} \\ 0 & 0 & 0 & 1 & -2 & 1 \end{bmatrix}$$

Figure 4.2: | The inverse from the reduced echelon form of an augmented matrix |

The matrices A and B should be entered and defined as in Solved Problem 1. Enter
1/4 *(A^(-2) &* (B &* A - A &* B) &* A + 3*B); (line 1, Figure 4.3). Note: Don't
forget the &* to indicate matrix multiplication, the * to indicate scalar multiplication,
and that the parentheses around -2 are necessary. (Refer to Section A.9 of Appendix
A to refresh your memory if needed.) If we evaluate line 1 of Figure 4.3 as a matrix,
we obtain the answer.

```
> 1/4 *(A^(-2) &* (B &* A - A &* B) &* A + 3*B);
```

$$\frac{1}{4}\left(\left(\frac{1}{A^2} \ \&^* \ ((B \ \&^* \ A) - (A \ \&^* \ B))\right) \ \&^* \ A\right) + \frac{3}{4}B$$

```
> evalm(");
```

$$\begin{bmatrix} \frac{3119}{28} & \frac{1782}{7} & \frac{10253}{28} \\ \frac{55}{28} & \frac{23}{28} & \frac{15}{28} \\ \frac{-773}{28} & \frac{-961}{14} & \frac{-2827}{28} \end{bmatrix}$$

Figure 4.3: | Solving an equation for a matrix |

Solved Problem 3: Use the inverse to solve the following system of equations.

$$\begin{aligned} 7x + 4y - 2z + 4w &= 8 \\ 2x - 3y + 7z - 6w &= 4 \\ 5x + 6y + 2z - 5w &= 2 \\ 3x + 3y - 5z + 8w &= 9 \end{aligned}$$

<u>Solution</u>: If $A = \begin{bmatrix} 7 & 4 & -2 & 4 \\ 2 & -3 & 7 & -6 \\ 5 & 6 & 2 & -5 \\ 3 & 3 & -5 & 8 \end{bmatrix}$, $\mathbf{u} = \begin{bmatrix} x \\ y \\ z \\ w \end{bmatrix}$, $\mathbf{v} = \begin{bmatrix} 8 \\ 4 \\ 2 \\ 9 \end{bmatrix}$, then the system of

equations is equivalent to the matrix equation $A\mathbf{u} = \mathbf{v}$. Multiplying each side of the equation on the left by A^{-1} produces the solution $\mathbf{u} = A^{-1}\mathbf{v}$. After the matrix A is defined and named, define the vector $\mathbf{v}$ by entering v := vector([8,4,2,9]). Now evaluate the expression A^(-1) &* v as a matrix. (<u>Note</u>: We did *not* use the transpose of $\mathbf{v}$ as you might see in your text. See Section A.8 of Appendix A for an explanation. Also, the spaces around the &* are necessary.) The solution $x = -\frac{29}{28}$, $y = \frac{21}{8}$, $z = \frac{295}{56}$, $w = \frac{107}{28}$ can now be read from line 3 of Figure 4.4.

```
> A := matrix(4,4, [7,4,-2,4, 2,-3,7,-6, 5,6,2,-5, 3,3,-5,8]);
```

$$A := \begin{bmatrix} 7 & 4 & -2 & 4 \\ 2 & -3 & 7 & -6 \\ 5 & 6 & 2 & -5 \\ 3 & 3 & -5 & 8 \end{bmatrix}$$

```
> v := vector([8,4,2,9]);
```

$$\mathbf{v} := \begin{bmatrix} 8 & 4 & 2 & 9 \end{bmatrix}$$

```
> evalm(A^(-1) &* v);
```

$$\begin{bmatrix} \dfrac{-29}{28} & \dfrac{21}{8} & \dfrac{295}{56} & \dfrac{107}{28} \end{bmatrix}$$

Figure 4.4: Using the inverse to solve a system of equations

4.3   Exercises

1. Use matrix inverses to solve the following systems of equations.

(a)
$$7x + 4y - 2z = 8$$
$$4x + 7y + 5z = 5$$
$$2x - 3y + 8z = 2$$

(b)
$$2x - 5y + z = a$$
$$7x - y + 4z = b$$
$$3x - 6y + 5z = 9$$

(c)
$$ax + 4y + 2z = 8$$
$$3x + by - 5z = 1$$
$$4x + 3y + cz = 3$$

2. For what values of s and t does $\begin{bmatrix} 2 & 7 & t \\ 3 & s & 1 \\ 2 & 4 & 6 \end{bmatrix}$ have an inverse?

3. Suppose that neither $\begin{bmatrix} 1 & 2 & s \\ 2 & 3 & t \\ 4 & 5 & 7 \end{bmatrix}$ nor $\begin{bmatrix} 4 & 5 & 8 \\ s & 2 & 3 \\ t & 1 & 8 \end{bmatrix}$ has an inverse. Find s and t.

4. Let $C = \begin{bmatrix} 3 & 4 & 1 \\ 9 & 2 & 7 \\ 4 & 1 & 5 \end{bmatrix}$ and $D = \begin{bmatrix} 5 & 6 & 2 \\ 9 & 4 & 7 \\ 2 & 2 & 1 \end{bmatrix}$. Solve the following equations for the 3 by 3 matrix A.

(a) $CAD = C - D$

(b) $C^2(A + 2D) = DA$

(c) $3AD^{-1} = A^2 DC$ (Assume that A is invertible.)

5. If A is a square matrix and $f(x) = p(x)/q(x)$, where p and q are polynomials, we define $f(A)$ to be $p(A)\, q(A)^{-1}$ provided that $q(A)^{-1}$ exists.

(a) If $A = \begin{bmatrix} 5 & 6 \\ 2 & 8 \end{bmatrix}$, evaluate the following rational functions at A.

(i) $\dfrac{x^3 + 2}{x^2 + 1}$

(ii) $\dfrac{x^2 + x - 2}{x - 1}$

(b) Show that $\dfrac{x^2 + x - 2}{x - 1} = x + 2$ if $x \neq 1$. Evaluate both sides for $x = A$. Are they the same?

(c) If $B = \begin{bmatrix} 29 & 45 \\ -18 & -28 \end{bmatrix}$, evaluate $\dfrac{x^2 + x - 2}{x - 1}$ and $x + 2$ at B. Are they the same? Explain what you observe.

6. Let $A = \begin{bmatrix} 1 & 3 & 2 \\ 4 & 5 & 1 \\ 3 & 7 & 2 \end{bmatrix}$. Find a quadratic polynomial $f(x)$ such that $f(A) = A^{-1}$.

(<u>Hint</u>: Find p, q, and r such that $pA^2 + qA + rI = A^{-1}$, where I is the 3 by 3 identity matrix. Remember that **convert** will change a matrix of equations to a set of equations that Maple can solve.)

7. In Solved Problem 4 of Chapter 3, we found a cubic polynomial $f(x)$ such that $f(A) = 0$. Show that the matrix A is invertible. Use the solution to Solved Problem 4 of Chapter 3 to express A^{-1} as a quadratic polynomial in A.

Prove that if a matrix is invertible and satisfies a polynomial equation $f(A) = 0$, then A^{-1} can be expressed as a polynomial in A. (Do not use computers here.)

8. **Markov chains revisited**: This problem refers to Exercise 9 in Chapter 3. By looking at powers of the transition matrix in that exercise, we could predict population trends. Powers of the inverse of the matrix tell us what happened in the past.

What would you expect to find if you looked up the population figures for the cities in 1985? in 1980?

4.4   Exploration and Discovery

1. Try to decide which of the following equations are true for matrices. For those that are, give a proof. For those that are not, give a counterexample. Use Maple to verify your counterexamples.

   (a) $(AB)^{-1} = A^{-1}B^{-1}$

   (b) $(A^3)^{-1} = (A^{-1})^3$

   (c) $(A^{-1}BA)^3 = A^{-1}B^3A$

   (d) $(ABC)^{-1} = C^{-1}B^{-1}A^{-1}$

   (e) Make up other formulas that are true for numbers but not matrices.

2. An upper triangular matrix is a matrix all of whose entries below the main diagonal are zero. For example, $\begin{bmatrix} 1 & 2 & 3 & 4 \\ 0 & 3 & 2 & 3 \\ 0 & 0 & 1 & 2 \\ 0 & 0 & 0 & 5 \end{bmatrix}$ is upper triangular. Calculate the inverses of several 4 by 4 upper triangular matrices. Is the inverse of an upper triangular matrix necessarily upper triangular? Prove your answer.

3. First do Exercise 5 where we defined $f(A)$ to be $p(A)\, q(A)^{-1}$.

   (a) Repeat the exercise using the definition $f(A) = q(A)^{-1}p(A)$. Compare your answers and discuss the results. Do you suspect $p(A)\, q(A)^{-1} = q(A)^{-1}p(A)$ for any square matrix A for which $q(A)$ is invertible? Can you prove it?

   (b) Now consider polynomials in two variables such as $p(x,y) = xy^2 + 3$ or $q(x,y) = 2x + y$. Calculate $p(A,B)$ and $q(A,B)$ when A and B are the matrices in Solved Problem 1.

   (c) Explore the two methods of defining $f(A,B)$ where $f(x,y) = p(x,y)/q(x,y)$ and discuss your findings.

4. $\boxed{\textbf{For students who have studied calculus.}}$ In Exploration and Discovery Problem 3 in Chapter 3, we discussed how one might define $\cos(A)$ and e^A, where A is a square matrix, by using the Maclaurin series. Since $1/\cos(x) = \sec(x)$ and $1/e^x = e^{-x}$, we may be led to conjecture that the inverses of the matrices

$\cos(A)$ and e^A are, respectively, $\sec(A)$ and e^{-A}. (How would you define $\sec(A)$ for a matrix A?) By working with the series, try to find evidence that the conjectures are true. Discuss your observations and findings.

4.5   Animal Migration

LABORATORY EXERCISE 4–1

Name _____   **Due Date** _____

Before beginning, it may be helpful to look at Exercise 8.

<u>The problem:</u> The state of Maine has been subdivided into four regions labeled A, B, C, and D to study migration patterns of a certain species of animal. In a given year some animals move from one region to another while some remain where they are. Studies have shown that animals from a given region move to other regions with probabilities given by the following table.

Probability of moving	From A	From B	From C	From D
to A	0.7	0.05	0.1	0.2
to B	0.1	0.60	0.2	0.1
to C	0.1	0.20	0.5	0.1
to D	0.1	0.15	0.2	0.6

In 1990 the populations of the regions were

$$A : 5500 \qquad B : 4800 \qquad C : 4100 \qquad D : 5250$$

For purposes of this exercise assume that these are the only factors governing the animal populations in these regions.

1. Let a_n denote the population of region A in year n, b_n the population of B in year n, c_n the population of C in year n, and d_n the population of D in year n.

   Let $\mathbf{p}_n$ denote the vector (a_n, b_n, c_n, d_n). If $A = \begin{bmatrix} 0.7 & 0.05 & 0.1 & 0.2 \\ 0.1 & 0.6 & 0.2 & 0.1 \\ 0.1 & 0.2 & 0.5 & 0.1 \\ 0.1 & 0.15 & 0.2 & 0.6 \end{bmatrix}$, show

   that $A\mathbf{p}_n = \mathbf{p}_{n+1}$.

2. Use a matrix equation to show how to get $\mathbf{p}_n$ from $\mathbf{p}_0$.

3. What are the expected populations of the four regions in 1995 and 2000?

4. Describe the eventual populations of the four regions. Explain how you arrived at your answer.

5. If you examined the regions in 1985, what population would you expect to find?

6. A prominent ecologist contends that this migration pattern has been persisting for at least 10 years. Criticize this statement.

Chapter 5
DETERMINANTS, ADJOINTS, AND CRAMER'S RULE

LINEAR ALGEBRA CONCEPTS

- **Determinant**
- **Adjoint**
- **Cramer's rule**

5.1 Introduction

The determinant is a function that assigns to each square matrix A a real number $\det(A)$. This number is important in determining if a matrix has an inverse and can be used to directly obtain solutions of certain systems of equations. Some books use $|A|$ for the determinant of the matrix A and some use $\det(A)$. Maple uses the latter syntax. (Exercise 1 asks you to compare the answers Maple gives for the two.)

5.2 Solved Problems

Solved Problem 1: Calculate the determinant and the adjoint of the matrix

$$A := \begin{bmatrix} 4 & 2 & -3 & 5 \\ 7 & 3 & 9 & 8 \\ 4 & 5 & 3 & 2 \\ 6 & 6 & 5 & 4 \end{bmatrix}$$

Solution: First, load the Linear Algebra package: `with(linalg)`. We won't be showing this in the figures from this point on. Enter and name the matrix A (line 1, Figure 5.1) as we have done in the previous chapters. Now enter `det(A)`. The answer is line 2 of Figure 5.1. To obtain the adjoint, we use `adjoint(A)`. The result appears in line 3 of Figure 5.1.

```
> A := matrix(4,4, [4,2,-3,5, 7,3,9,8, 4,5,3,2, 6,6,5,4]);
```

$$A := \begin{bmatrix} 4 & 2 & -3 & 5 \\ 7 & 3 & 9 & 8 \\ 4 & 5 & 3 & 2 \\ 6 & 6 & 5 & 4 \end{bmatrix}$$

```
> det(A);
```

$$39$$

```
> adjoint(A);
```

$$\begin{bmatrix} -10 & -63 & -311 & 294 \\ 6 & 30 & 171 & -153 \\ -4 & 6 & 16 & -15 \\ 11 & 42 & 190 & -183 \end{bmatrix}$$

Figure 5.1: | The determinant and adjoint of a matrix |

Solved Problem 2: For the matrix in Solved Problem 1, verify that if any multiple of row 2 is added to row 1, the value of the determinant does not change.

<u>Solution</u>: Enter the matrix

$$\begin{bmatrix} 4+7t & 2+3t & -3+9t & 5+8t \\ 7 & 3 & 9 & 8 \\ 4 & 5 & 3 & 2 \\ 6 & 6 & 5 & 4 \end{bmatrix}$$

(Alternatively, you may use the `addrow` function to perform this row operation. See item 5 of Section A.9 in Appendix A .)

Now enter `det('')`. The result in line 2 of Figure 5.2 is the same as the determinant of the original matrix appearing in Figure 5.1 and is independent of t.

```
> matrix(4,4, [4+7*t,2+3*t,-3+9*t,5+8*t, 7,3,9,8, 4,5,3,2, 6,6,5,4]);
```

$$\begin{bmatrix} 4+7t & 2+3t & -3+9t & 5+8t \\ 7 & 3 & 9 & 8 \\ 4 & 5 & 3 & 2 \\ 6 & 6 & 5 & 4 \end{bmatrix}$$

```
> det(");
```

$$39$$

Figure 5.2: | Effect of an elementary row operation on a matrix |

Solved Problem 3: Use Cramer's rule to solve the following system of equations for z.

$$\begin{aligned} 4x + 7y - 8z + 2w &= 9 \\ 3x - y + 2z + 9w &= 7 \\ 5x + 6y + 2z - w &= 3 \\ 8x - 3y + 2z - w &= 5 \end{aligned}$$

Solution: Enter and name the coefficient matrix $A := \begin{bmatrix} 4 & 7 & -8 & 2 \\ 3 & -1 & 2 & 9 \\ 5 & 6 & 2 & -1 \\ 8 & -3 & 2 & -1 \end{bmatrix}$ and the

vector of contants $\mathbf{b} := \begin{bmatrix} 9 \\ 7 \\ 3 \\ 5 \end{bmatrix}$. It is easy to build the numerator matrix needed

by Cramer's Rule to solve for z by using Maple's augment in concert with the col selection function. The purpose of col is to copy columns from a matrix. The syntax for col has two forms:

$$\text{col}(matrix,\ column)$$
$$\text{col}(matrix,\ start..end)$$

The first form selects a single column from *matrix*, the second copies all columns from *start* to *end*. We need both forms. To form the numerator, use augment(col(A,1..2), b, col(A,4)). Next, find det('')/det(A). The results are in Figure 5.3.

```
> A := matrix(4,4, [4,7,-8,2, 3,-1,2,9, 5,6,2,-1, 8,-3,2,-1]);
```

$$A := \begin{bmatrix} 4 & 7 & -8 & 2 \\ 3 & -1 & 2 & 9 \\ 5 & 6 & 2 & -1 \\ 8 & -3 & 2 & -1 \end{bmatrix}$$

```
> b := vector([9,7,3,5]);
```

$$b := \begin{bmatrix} 9 & 7 & 3 & 5 \end{bmatrix}$$

```
> augment(col(A,1..2), b, col(A,4));
```

$$\begin{bmatrix} 4 & 7 & 9 & 2 \\ 3 & -1 & 7 & 9 \\ 5 & 6 & 3 & -1 \\ 8 & -3 & 5 & -1 \end{bmatrix}$$

```
> det(")/det(A);
```

$$\frac{-2983}{6024}$$

Figure 5.3: Solving a system of equations using Cramer's rule

5.3   Exercises

1. Enter the matrix $A := \begin{bmatrix} a & b \\ c & d \end{bmatrix}$.

   (a) Some books use $|A|$ for the determinant of the matrix A and some use $\det(A)$. What do you get if you ask Maple to simplify $|A|$? That is, what do you get if you ask Maple to evalm(abs(A))?

   (b) For students who have studied calculus. Some books use $|\mathbf{v}|$ for the norm (magnitude or length) of the vector $\mathbf{v}$ and some use norm($\mathbf{v}$). What do you get if you ask Maple to simplify norm(A, frobenius)? (The name "frobenius" indicates the usual Euclidean distance measure; see the help page for norm for further information.) What do you get if you ask Maple to evaluate norm(A, frobenius)?

   Give an example of a *nonzero* 2 by 2 matrix A such that Maple gives the same answer for norm(A, frobenius) and $\det(A)$.

2. If $A = \begin{bmatrix} 2 & 3 & 5 & 8 \\ 1 & 3 & 2 & 3 \\ 7 & 5 & 4 & 1 \\ 9 & 8 & 1 & 2 \end{bmatrix}$ and $B = \begin{bmatrix} 3 & 6 & 4 & 2 \\ 5 & 4 & 2 & 7 \\ 3 & 9 & 4 & 1 \\ 2 & 4 & 6 & 8 \end{bmatrix}$, calculate the determinants of A, B, and each of the following.

(i)	AB	(ii)	BA	(iii)	$AB - BA$
(iv)	$(AB)^2$	(v)	A^2B^2	(vi)	$(AB)^2 - A^2B^2$

   (a) Given your answers to (i) and (ii), explain why the answer to (iii) is not zero.

   (b) Given your answers to (iv) and (v), explain why the answer to (vi) is not zero.

3. Calculate the adjoints of the matrices in Exercise 2 above.

4. Use the determinant to determine the values of t for which the following system of equations has a unique solution.

$$
\begin{aligned}
2x + 4y + t\,z &= 3 \\
t\,x - y + z &= 8 \\
3x + t\,y - z &= 12
\end{aligned}
$$

5. Find a multiple of the matrix $\begin{bmatrix} 3 & 2 & 9 & 5 \\ 5 & 3 & 7 & 9 \\ 1 & 3 & 4 & 6 \\ 4 & 2 & 3 & 1 \end{bmatrix}$ that has determinant 1.

6. In Chapter 3 we discussed how, given a matrix M, to find a polynomial $p(x)$ such that $p(M) = 0$.

   (a) If $M = \begin{bmatrix} 3 & 2 \\ 6 & 1 \end{bmatrix}$, find $\det(xI - M)$. (Defining `Id := &*()` will help. Also recall that `sort(p,x)` and `collect(p,x)` are useful with polynomials.)

   (b) Your answer above was a polynomial. Evaluate it for $x = M$.

   (c) Repeat parts (a) and (b) for the matrix in Solved Problem 4 of Chapter 3. Compare the answer you obtain here with that of Solved Problem 4.

7. Repeat parts (a) and (b) of Exercise 6 above for the matrix $M = \begin{bmatrix} a & b \\ c & d \end{bmatrix}$.

   <u>Remark</u>: It may seem obvious that if $p(x) = \det(xI - M)$, then $p(M) = \det(MI - M) = \det(0) = 0$. This is a correct result but not a correct argument. See Exploration and Discovery Problem 2(b).

8. Repeat parts (a) and (b) of Exercise 6 above for the matrix $M = \begin{bmatrix} a & b & c \\ d & e & f \\ g & h & i \end{bmatrix}$.

   See the hints given in Exercise 7.

9. Use Cramer's rule to solve the following system of equations for b.

$$\begin{aligned} 3a - 4b + 2c - d + 5e &= 3 \\ 8a + 7b - 9c + 2d - e &= 5 \\ 6a - 2b + 3c - 3d + 4e &= 6 \\ 5a + 4b + 6c + 8d + e &= 2 \\ 8a + 2b - 5c + 5d - 9e &= 7 \end{aligned}$$

10. Solve the system in the preceding problem and verify that the solution for b is the same as that given by Cramer's rule.

5.4 Exploration and Discovery

1. For each of the following equations, if it is true, prove it. If it is not, find a counterexample. Assume that A and B are square matrices in each case.

   (a) $\det(A + B) = \det(A) + \det(B)$

   (b) $\det(AB) = \det(A)\det(B)$

   (c) $\det(A^n) = \det(A)^n$

   (d) If A is invertible, then $\det(A^{-1}) = 1/\det(A)$.

   (e) $\det(cA) = c\det(A)$ for any real number c.

   (f) $\det(A B A^{-1}) = \det(B)$

   (g) Let A be an n by n matrix and let B denote the adjoint of A. Then $\det(B) = (\det(A))^{n-1}$.

2. First you should do Exercises 7 and 8.

   (a) In those two problems we calculated $p(x) = \det(xI - M)$, evaluated it for $x = M$, and got zero. This important fact is called the *Cayley-Hamilton Theorem*. Verify this theorem for 4 by 4 matrices by repeating the procedure in Exercises 7 and 8.

   (b) The Cayley-Hamilton Theorem may seem obvious because of the following "proof."

   $$p(M) = \det(M I - M) = \det(\text{ zero matrix }) = 0$$

   However, this argument is *not* correct; explain why. Here is a clue: If M is a 2 by 2 matrix, what is $p(M)$ (a number, vector, matrix, ... ?), then what is $\det(\text{ zero matrix })$ (a number, vector, matrix, ... ?)

   (c) Do Laboratory Exercise 3.1. It examines $\det(A + xB)$ for two choices of B. It is also a polynomial in x. For each choice of B, try to discover a matrix C and a constant k such that $\det(A + xB) = k\det(xI - C)$. (<u>Hint</u>: Use paper and pencil.) Report your observations and conclusions.

3. Find four 3 by 3 matrices with integer entries (other than triangular matrices) that have determinant 1 or -1. Explain how you arrived at your answers and discuss a strategy for generating such matrices. (<u>Hint</u>: Try letting one or two entries be variables and the rest be numbers.)

4. The matrices in (a) and (c) below are known as *Vandermonde matrices*.

   (a) Calculate det $\left(\begin{bmatrix} 1 & x & x^2 \\ 1 & y & y^2 \\ 1 & z & z^2 \end{bmatrix} \right)$ and **factor** your answer.

   (b) Based on your answer above, give a necessary and sufficient condition for this determinant to be zero.

   (c) Repeat (a) and (b) for $\begin{bmatrix} 1 & x & x^2 & x^3 \\ 1 & y & y^2 & y^3 \\ 1 & z & z^2 & z^3 \\ 1 & w & w^2 & w^3 \end{bmatrix}$.

   (d) Conjecture a factorization for larger Vandermonde matrices and give a necessary and sufficient condition for their determinants to be zero.

5. $\begin{bmatrix} 1 & \frac{1}{2} \\ \frac{1}{2} & \frac{1}{3} \end{bmatrix}$ and $\begin{bmatrix} 1 & \frac{1}{2} & \frac{1}{3} \\ \frac{1}{2} & \frac{1}{3} & \frac{1}{4} \\ \frac{1}{3} & \frac{1}{4} & \frac{1}{5} \end{bmatrix}$ are known as *Hilbert* matrices of orders 2 and 3 respectively. To assist in entering a large Hilbert matrix, Maple has a function **hilbert**. If you now enter **hilbert(n)**, Maple will return the Hilbert matrix of order n.

   (a) Find the determinants of the Hilbert matrices of orders 2, 3, 5, and 10.

   (b) What do you observe about the magnitude of the determinants of Hilbert matrices?

   (c) Based on your calculations make a conjecture concerning the invertibility of Hilbert matrices. (Try to prove your conjecture.)

   (d) There are many computer programs written in languages such as C, BASIC, Pascal, or Fortran that find inverses and determinants. Unlike Maple, such programs use "floating point" decimal approximations (such as 0.33333333 for $\frac{1}{3}$), which introduces a small error. What do you think will happen if you use one of these programs to calculate the determinant of a large Hilbert matrix? How about the inverse? If you are able to get access to one, try the 10 by 10 example and report what happens. Try it with Maple in approximate mode by using **map** to apply **evalf** to the fractions in the matrix as in **map(evalf, A)** after setting Maple's precision to 5 digits with **Digits:=5**.

6. Calculate the determinant of $A = \begin{bmatrix} 2 & 3 & 5 & 2 \\ 1 & 2 & 1 & 3 \\ 2 & 4 & 1 & 5 \\ 3 & 1 & 2 & 3 \end{bmatrix}$. Change any entry of A by

$\frac{1}{100}$ and calculate the determinant again. For example, try $\begin{bmatrix} 2 & 3 & 5 & 2.01 \\ 1 & 2 & 1 & 3 \\ 2 & 4 & 1 & 5 \\ 3 & 1 & 2 & 3 \end{bmatrix}$.

How much does the determinant change? Perform this same experiment with other square matrices altering several of the entries by a small amount. Discuss your observations.

Repeat the preceding experiment with the noninvertible matrix $A = \begin{bmatrix} 1 & 2 & 3 \\ 4 & 5 & 6 \\ 7 & 8 & 9 \end{bmatrix}$.

If you select the entries of a square matrix at random, do you expect its determinant to be 0? Do you expect it to be invertible? (Try several using `randmatrix`. See Appendix B for information about `randmatrix`.)

7. | **Continuation of Problem 6 for students who have studied calculus.** |
Intuitively, a function f is *continuous* if a small change in x produces only a small change in $f(x)$. Discuss the continuity of the determinant as a function of one of its entries.

5.5   The Determinant of a Sum

LABORATORY EXERCISE 5–1

Name _____   **Due Date** _____

The <u>problem</u>: If A and B are n by n matrices and x is a number, we know that the equations $\det(A + B) = \det(A) + \det(B)$ and $\det(A + xB) = \det(A) + x\det(B)$ are usually false. But if A, B, and x are chosen carefully, we can find cases where these are true. We will investigate two examples. Write your responses to all problems in clear, grammatically correct English sentences.

Let $A = \begin{bmatrix} 3 & 0 & 1 \\ 6 & 1 & 1 \\ 9 & 0 & 7 \end{bmatrix}$, let Id be the 3 by 3 identity matrix, and let $B = \begin{bmatrix} 1 & 2 & 1 \\ 3 & 4 & 3 \\ -4 & -2 & -4 \end{bmatrix}$.

1. Find $\det(A)$, $\det(B)$, $\det(Id)$, $\det(A + xId)$, and $\det(A + xB)$.

2. Find all values of x for which $\det(A + xId) = \det(A) + \det(xId)$.

3. Find all values of x for which $\det(A + xB) = \det(A) + \det(xB)$.

4. Compare and contrast your results in the last two problems.

continued on next page

5. Find all values of x for which $\det(A + xId) = \det(A) + x\det(Id)$.

6. Find all values of x for which $\det(A + xB) = \det(A) + x\det(B)$.

7. Compare and contrast your results in the last two problems.

8. Find all values of x such that $A + xId$ is *not* invertible.

9. Find all values of x such that $A + xB$ is *not* invertible.

10. Compare and contrast your results in the last two problems.

Chapter 6
APPLICATION: MATRIX ALGEBRA AND MODULAR ARITHMETIC

LINEAR ALGEBRA CONCEPTS

- **Modular arithmetic**
- **Matrix operations**
- **Hill codes**

6.1 Introduction

This chapter expands the usual matrix operations using arithmetic modulo a prime. This material is not covered in most linear algebra texts, and it can be considered an optional chapter. However, the laboratory exercise at the end of this chapter contains an interesting application to cryptography called Hill codes*.

6.2 A Brief Review of Modular Arithmetic

Let $\mathbb{Z}_n$ denote the set $\{0, 1, 2, \ldots, n-1\}$. If k is any integer, the residue of k modulo n is the remainder of k divided by n. If the remainder is r we write $k = r(\text{mod } n)$. For example, $29 = 4(\text{mod } 5)$ because when 29 is divided by 5 we get a remainder of 4. Addition and multiplication in $\mathbb{Z}_n$ have the usual definitions except that if the sum or product gets too large, we replace it by its residue modulo n. Thus, in $\mathbb{Z}_5$, $2 \cdot 3 = 6(\text{mod } 5) = 1$ and $3 + 4 = 7(\text{mod } 5) = 2$.

Maple's syntax for the residue of k mod n is k mod n or `mod`(k,n) (note the backquotes around mod). The latter form will be more convenient for matrix calculations. Thus, if we enter `mod`(17,5), Maple will return the value 2 because this is the remainder of 17 divided by 5. Maple will also perform this operation on matrices, but we must do it element by element via the map command. If A

---

*These codes are based on Hill's 1929 and 1931 papers "Cryptography in an algebraic alphabet," *Amer. Math. Monthly* **36**, 306–312, and "Concerning certain linear transformation apparatus of cryptography," *Amer. Math. Monthly* **38**, 135–154.

is the matrix $\begin{bmatrix} 0 & 1 & 2 \\ 3 & 4 & 5 \\ 6 & 7 & 8 \end{bmatrix}$ and we enter `map('mod', A, 5)`, then Maple will return

$\begin{bmatrix} 0 & 1 & 2 \\ 3 & 4 & 0 \\ 1 & 2 & 3 \end{bmatrix}$. It will be helpful to define the function `matmod` by

$$\texttt{matmod := (M,p) -> map('mod', evalm(M), p);}$$

The statement above becomes `matmod(A, 5)`.

Division modulo an integer n can cause difficulties. For example, the obvious solution of the equation $3x = 8(\text{mod } 13)$ is $x = \frac{8}{3}$. But this is not satisfactory because $\frac{8}{3}$ is not an integer in $\{0, 1, \cdots, 12\}$. The following theorem due to the sixteenth-century mathematician Leonhard Euler is needed. Its proof can be found in any elementary number theory text.

Theorem (Euler): If p is prime and k is not a multiple of p, then $k^{p-1} = 1(\text{mod } p)$.

The theorem tells us that $3^{12} = 1(\text{mod } 13)$. Maple can check this: enter `mod(3^12, 13)`. Thus, $\frac{8}{3} = (\frac{8}{3})(1) = (\frac{8}{3})3^{12}(\text{mod } 13)$. If we enter `mod(8/3 * 3^12, 13)` we find that $\frac{8}{3} = 7(\text{mod } 13)$.

In summary, to solve an equation $ax = b(\text{mod } p)$, calculate the answer $x = \frac{b}{a}$ as a rational number and use Euler's theorem to clear the denominator.

6.3   Solved Problems

<u>Solved Problem 1</u>: Solve the following system of equations.

$$\begin{aligned} 3x + 5y - 7z &= 8\,(\text{mod } 83) \\ 8x - 9y + 13z &= 13\,(\text{mod } 83) \\ 7x + 4y + 5z &= 15\,(\text{mod } 83) \end{aligned}$$

<u>Solution</u>: Enter the augmented matrix of the system of equations and row-reduce. We need to clear the fractions from line 2 of Figure 6.1. Before applying Euler's theorem, you should check that 83 is not a divisor of 701. After this fact has been verified, enter `matmod(701^82 * '', 83)`. (Since $701^{82} = 1\,(\text{mod } 83)$, this multiplication does not change the matrix mod 83.) The solution, $x = 50\,(\text{mod } 83)$, $y = 5\,(\text{mod } 83)$, and $z = 12\,(\text{mod } 83)$, can be read from the last column of line 3.

```
> A := matrix(3,4, [3,5,-7,8, 8,-9,13,13, 7,4,5,15]);
```

$$A := \begin{bmatrix} 3 & 5 & -7 & 8 \\ 8 & -9 & 13 & 13 \\ 7 & 4 & 5 & 15 \end{bmatrix}$$

```
> rref(A);
```

$$\begin{bmatrix} 1 & 0 & 0 & \frac{1435}{701} \\ 0 & 1 & 0 & \frac{185}{701} \\ 0 & 0 & 1 & \frac{-54}{701} \end{bmatrix}$$

```
> matmod(701^82 * A, 83);
```

$$\begin{bmatrix} 1 & 0 & 0 & 50 \\ 0 & 1 & 0 & 5 \\ 0 & 0 & 1 & 12 \end{bmatrix}$$

Figure 6.1: | Solving a system of equations modulo 83 |

Solved Problem 2: Find the inverse of $A = \begin{bmatrix} 2 & 17 & 28 \\ 8 & 19 & 2 \\ 21 & 4 & 16 \end{bmatrix}$ modulo 29.

Solution: In this context, the familiar theorem that A has an inverse provided $\det(A) \neq 0$ must be replaced by the theorem

A has an inverse modulo a prime p if and only if p does not divide $\det(A)$

Enter the matrix and name it A. Ask Maple for det(A). Since the determinant, -11146, is not divisible by 29, we conclude that A has an inverse modulo 29.

The next step is to find A^{-1} (line 2, Figure 6.2). Because $2 \cdot 5573 = 11146$, we can clear the fractions in A^{-1} by multiplying by any power of 11146. Euler's theorem tells us that $11146^{28} = 1 \pmod{29}$. Thus, we enter matmod(11146^28 * ", 29). The final result is displayed in line 3. You may wish to check this answer using matmod(A &* ", 29).

```
> det(A);
```
$$-11146$$

```
> evalm(A^(-1));
```

$$\begin{bmatrix} \dfrac{-148}{5573} & \dfrac{80}{5573} & \dfrac{249}{5573} \\[2mm] \dfrac{43}{5573} & \dfrac{278}{5573} & \dfrac{-110}{5573} \\[2mm] \dfrac{367}{11146} & \dfrac{-349}{11146} & \dfrac{49}{5573} \end{bmatrix}$$

```
> matmod(11146^28 * ", 29);
```

$$\begin{bmatrix} 11 & 16 & 15 \\ 26 & 15 & 7 \\ 28 & 26 & 4 \end{bmatrix}$$

Figure 6.2: | The inverse of a matrix modulo 29 |

Solved Problem 3: A is a 3 by 3 matrix with integer entries such that $A \begin{bmatrix} 2 \\ 5 \\ 8 \end{bmatrix} = \begin{bmatrix} 3 \\ 9 \\ 2 \end{bmatrix}$ (mod 29), $A \begin{bmatrix} 4 \\ 6 \\ 1 \end{bmatrix} = \begin{bmatrix} 5 \\ 9 \\ 4 \end{bmatrix}$ (mod 29), and $A \begin{bmatrix} 8 \\ 7 \\ 6 \end{bmatrix} = \begin{bmatrix} 4 \\ 5 \\ 2 \end{bmatrix}$ (mod 29). Find $A \begin{bmatrix} 6 \\ 5 \\ 8 \end{bmatrix}$ (mod 29).

Solution: Let $C = \begin{bmatrix} 2 & 4 & 8 \\ 5 & 6 & 7 \\ 8 & 1 & 6 \end{bmatrix}$ and $B = \begin{bmatrix} 3 & 5 & 4 \\ 9 & 9 & 5 \\ 2 & 4 & 2 \end{bmatrix}$. The three equations above are equivalent to the single matrix equation $AC = B \,(\text{mod } 29)$. Thus, if C has an inverse modulo 29, then $A = B C^{-1}(\text{mod } 29)$. Enter these matrices, naming them C and B. If you enter `det(C)`, Maple will return -182. This is not divisible by 29, and we conclude that C has an inverse modulo 29. Enter, evaluating as a matrix expression, `B &* C^(-1)`. From the result in line 1 of Figure 6.3, we need to multiply by a power of 182 to clear the fractions. Once again applying Euler's theorem, enter `A := matmod(182^28 * '', 29)`. The matrix A is the result displayed in line 2 of Figure 6.3.

Finally, enter `matmod(A &* vector([6,5,8]), 29)`. The answer appears in line 3 of Figure 6.3.

```
> evalm(B &* C^(-1));
```

$$\begin{bmatrix} \frac{-45}{182} & \frac{94}{91} & \frac{-19}{91} \\ \frac{-20}{13} & \frac{33}{13} & \frac{-1}{13} \\ \frac{-38}{91} & \frac{90}{91} & \frac{-24}{91} \end{bmatrix}$$

```
> A := matmod(11146^28 * ", 29);
```

$$A := \begin{bmatrix} 27 & 9 & 17 \\ 23 & 7 & 20 \\ 5 & 8 & 23 \end{bmatrix}$$

```
> matmod(A &* vector([6,5,8]), 29);
```

$$\begin{bmatrix} 24 & 14 & 22 \end{bmatrix}$$

Figure 6.3: │ A matrix with given vector products │

6.4   Exercises

1. Solve the following systems of equations modulo 881.

   (a) $7x + 4y - 9z = 18$ (b) $14x + 13y + 6z - 4w = 17$
   $3x - 13y + 27z = 7$ $12x - 7y + 18z + 4w = 64$
   $14x - 77y + 38z = 93$ $8x + 4y - 15z + 18w = 9$

2. For each of the following matrices determine if the inverse modulo 29 exists. If it does, calculate it and check your answer.

   (a) $\begin{bmatrix} 13 & 4 & 7 \\ 18 & 2 & 9 \\ 3 & 14 & 6 \end{bmatrix}$ (b) $\begin{bmatrix} 8 & 7 & 6 \\ 12 & 4 & 11 \\ 2 & 9 & 7 \end{bmatrix}$

3. Suppose A is a 3 by 3 matrix with integer entries, such that

$$A \begin{bmatrix} 4 \\ 7 \\ 17 \end{bmatrix} = \begin{bmatrix} 2 \\ 20 \\ 13 \end{bmatrix} \pmod{29}, \quad A \begin{bmatrix} 3 \\ 16 \\ 4 \end{bmatrix} = \begin{bmatrix} 2 \\ 18 \\ 9 \end{bmatrix} \pmod{29}$$

$$A \begin{bmatrix} 5 \\ 3 \\ 8 \end{bmatrix} = \begin{bmatrix} 2 \\ 17 \\ 23 \end{bmatrix} \pmod{29}$$

   Find $A \pmod{29}$.

4. Let $A = \begin{bmatrix} 3 & 2 & 5 \\ 4 & 8 & 1 \\ 5 & 4 & 6 \end{bmatrix}$. Find an n such that $A^n = I \pmod{31}$. (<u>Hint</u>: The command `seq(matmod(A^n, 31), n=k..j)` will let you view the kth through the jth powers of A modulo 31.)

5. If A is an integer matrix and $A^{-1} = \begin{bmatrix} 4 & 6 & 9 \\ 2 & 1 & 4 \\ 19 & 3 & 6 \end{bmatrix} \pmod{37}$, then find $A \pmod{37}$.

6.5   Exploration and Discovery

1. If A is an integer matrix that has an inverse modulo a prime p, is there always an n such that $A^n = I \pmod{p}$? Look at several examples. (<u>Hint</u>: It may help to count how many matrices there are with entries from $\mathbb{Z}_p$.)

2. It can be shown that even if n is not prime, an integer matrix A still has an inverse modulo n provided n and $\det(A)$ have no common divisors other than 1. That is, $\gcd(n, \det(A)) = 1$. However, there is a difficulty in calculating the inverse since Euler's theorem is not true if n is not prime.

   (a) Show that $8^{20} \neq 1 \pmod{21}$.

   (b) Maple uses the syntax $\gcd(a, b)$ to denote the greatest common divisor of a and b. Find the greatest common divisors of the pairs $(273, 1491)$, and $(2186, 3294)$.

   (c) The Euler function $\phi(n)$ is defined to be the number of positive integers $k < n$ such that $\gcd(k, n) = 1$. For example, $\gcd(1, 6) = 1$, $\gcd(2, 6) = 2$, $\gcd(3, 6) = 3$, $\gcd(4, 6) = 2$, and $\gcd(5, 6) = 1$. Thus $\phi(6) = 2$. Calculate $\phi(26)$ and $\phi(883)$. (<u>Hint</u>: 883 is prime.)

   (d) A second theorem of Euler states

   $$\text{If } \gcd(a, n) = 1, \text{ then } a^{\phi(n)} = 1 \pmod{n}.$$

   Verify this theorem for $n = 26$ and $a = 17$.

   (e) Find the inverse of $\begin{bmatrix} 2 & 5 & 3 \\ 1 & 4 & 7 \\ 2 & 9 & 2 \end{bmatrix}$ modulo 26.

6.6 Hill Codes

LABORATORY EXERCISE 6.1

Name _____   Due Date _____

One common method of coding is the *Hill code*. (For a more detailed discussion, see *Elementary Linear Algebra, Applications Version* by Howard Anton and Chris Rorres.) The idea is as follows. We add the three symbols , . ? to the alphabet to make the total number prime, and each symbol is assigned a number according to the following table.

A	B	C	D	E	F	G	H	I	J	K	L	M	N	O
0	1	2	3	4	5	6	7	8	9	10	11	12	13	14

P	Q	R	S	T	U	V	W	X	Y	Z	,	.	?
15	16	17	18	19	20	21	22	23	24	25	26	27	28

To illustrate how it works, we will encode **HIDE THIS MESSAGE** using a Hill 3-code with *encryption matrix* $M = \begin{bmatrix} 2 & 7 & 6 \\ 4 & 5 & 13 \\ 2 & 6 & 1 \end{bmatrix}$.

Step 1: Remove the spaces and separate the letters into groups of 3. (Add arbitrary letters to the end if necessary.)

HID ETH ISM ESS AGE

Step 2: Assign each letter its appropriate number from the table above, and arrange these numbers in the columns of a so-called *plaintext* matrix, P as shown:

$$P = \begin{bmatrix} 7 & 4 & 8 & 4 & 0 \\ 8 & 19 & 18 & 18 & 6 \\ 3 & 7 & 12 & 18 & 4 \end{bmatrix}$$

Step 3: Calculate the product $M\,P$ modulo 29 to obtain the *code matrix*:

$$C = \begin{bmatrix} 1 & 9 & 11 & 10 & 8 \\ 20 & 28 & 17 & 21 & 24 \\ 7 & 13 & 20 & 18 & 11 \end{bmatrix}$$

Step 4: Replace the numbers in the code matrix, C by their corresponding letters.

$$\begin{bmatrix} B & J & L & K & I \\ U & ? & R & V & Y \\ H & N & U & S & L \end{bmatrix}$$

Step 5: Arrange the columns into a single line (without spaces) and send the text:

BUHJ?NLRUKVSIYL

Even if the message is intercepted by an unauthorized person, she will not be able to decode the message unless she has the order of the code (3) and the encryption matrix M. The person to whom we are sending the message has both and can decipher the message by repeating the process above using the inverse modulo 29 of the encryption matrix. Try the following exercises.

1. Calculate the inverse of the encryption matrix M, modulo 29.

2. Use your answer to decode the message **BUHJ?NLRUKVSIYL** that we just encoded above. Show your work.

3. Decode **TMB,.PDHJPHV** assuming that it is an order 3 Hill code with the encryption matrix above.

4. You intercept the following message: **GHTIOTZVEKVVXTVB,IZCX**. Naval Intelligence suspects that this is an order 3 Hill code and that the first three words of the message are **I THINK OUR**. Assuming this, find the encryption matrix and decode the message.

Chapter 7
VECTOR PRODUCTS, LINES, AND PLANES

LINEAR ALGEBRA CONCEPTS

- **Dot product**
- **Cross product**
- **Projection**
- **Unit vector**
- **Vectors in $\mathbb{R}^n$**
- **Orthogonal vectors**

7.1   Introduction

This chapter covers basic operations on vectors in the plane and in 3-space emphasizing the geometric aspects of linear algebra. Maple's syntax for the *dot product* of two vectors $\mathbf{x}$ and $\mathbf{y}$ is $\texttt{dotprod}(\mathbf{x}, \mathbf{y})$, and for *cross product* is $\texttt{crossprod}(\mathbf{x}, \mathbf{y})$. The *length* or *norm* (Euclidean) of a vector $\mathbf{x}$ is denoted in Maple by $\texttt{norm}(\mathbf{x}, 2)$ or $\texttt{norm}(\mathbf{x}, \texttt{frobenius})$. However, we shall define our own length function $\texttt{vlength}$ in terms of dot products since Maple's $\texttt{norm}$ uses absolute values that cause problems for $\texttt{solve}$. Since Maple's $\texttt{solve}$ does not handle matrix equations, we will also define a function $\texttt{matsolve}$. (See Appendix B.) After loading the $\texttt{linalg}$ package, define the functions:

```
vlength := x -> sqrt(dotprod(x,x))
matsolve := eq -> solve(convert(evalm(lhs(eq)-rhs(eq)), set))
```

7.2   Solved Problems

<u>Solved Problem 1</u>: If $\mathbf{u} = (1, 2, 3)$ and $\mathbf{v} = (2, 5, 7)$, find the following.

1. A unit vector in the direction of $\mathbf{u}$.

2. $\mathbf{u} \cdot \mathbf{v}$.

3. $\mathbf{u} \times \mathbf{v}$.

4. The projection of $\mathbf{u}$ onto $\mathbf{v}$.

<u>Solution</u>: Each of the items above is easily calculated by hand. The point of this exercise is to introduce Maple's syntax for these operations, which will be used later. The first step is to define the vectors; enter u := vector([1,2,3]) and v := vector([2,5,7]). (We entered v in the same input cell by pressing ⎡*shift return*⎤ after u to go to the next screen line without executing the command. Then pressing ⎡*return*⎤ after v executes both lines. Check the documentation on how to do this on your system.) To find a unit vector in the direction of $\mathbf{u}$, type u/vlength(u). Maple calculates the length, but doesn't evaluate fully as a vector; enter evalm('') to finish (line 3 of Figure 7.1). Enter dotprod(u,v) to get $\mathbf{u} \cdot \mathbf{v}$ (line 4). Enter crossprod(u,v) to get $\mathbf{u} \times \mathbf{v}$ (line 5). For the projection of $\mathbf{u}$ onto $\mathbf{v}$, enter evalm(dotprod(u,v)/dotprod(v,v)*v). The result is in line 6 of Figure 7.1.

```
>  u := vector([1,2,3]);
   v := vector([2,5,7]);
```
$$u := [1\ 2\ 3]$$
$$v := [2\ 5\ 7]$$

```
> u/vlength(u);
```
$$\tfrac{1}{14}\, u\, \sqrt{14}$$

```
> evalm(");
```
$$\left[\frac{1}{14}\sqrt{14}\quad \frac{1}{7}\sqrt{14}\quad \frac{3}{14}\sqrt{14}\right]$$

```
> dotprod(u, v);
```
$$33$$

```
> crossprod(u, v);
```
$$[-1\ -1\ \ 1]$$

```
> evalm( dotprod(u,v)/ dotprod(v,v) * v );
```
$$\left[\frac{11}{13}\quad \frac{55}{26}\quad \frac{77}{26}\right]$$

Figure 7.1: ⎡Basic operations on vectors⎤

Solved Problem 2: Find all unit vectors in $\mathbb{R}^3$ that make an angle of 1 radian with both $\mathbf{u} = (1,2,3)$ and $\mathbf{v} = (2,1,1)$.

<u>Solution</u>: Define $\mathbf{u} := (1,2,3)$, $\mathbf{v} := (2,1,1)$, and $\mathbf{w} := (a,b,c)$ as seen in Figure 7.2. Assuming $\mathbf{w}$ is a unit vector, the cosine of the angle between $\mathbf{w}$ and $\mathbf{u}$ is $(\mathbf{u} \cdot \mathbf{w}) \,/\, |\mathbf{u}|$. Enter the set $\{(u \cdot w)\,/\,|u| = \cos(1),\ (v \cdot w)\,/\,|v| = \cos(1)\}$ as seen in line 2. Now solve using *solve variables* a and b, and name the solution $\mathtt{soln}$. We can collect like terms in the result of line 4 to obtain the following representation for $\mathbf{w}$.

$$\mathbf{w} = \left(\frac{(2\sqrt{6} - \sqrt{14})\cos(1) + c}{3} \quad \frac{(2\sqrt{14} - \sqrt{6})\cos(1) - 5c}{3} \quad c \right)$$

We named $\mathtt{wc}$ in line 4 to avoid typing this difficult expression. <u>Note</u>: we must use $\mathtt{evalm}$ in line 4 to substitute into the elements of $\mathbf{w}$ rather than into the name "$\mathbf{w}$".

```
> u := vector([1,2,3]);
  v := vector([2,1,1]);
  w := vector([a,b,c]);
```

$$\begin{aligned} u &:= [1\ 2\ 3] \\ v &:= [2\ 1\ 1] \\ w &:= [a\ b\ c] \end{aligned}$$

```
> eqs := {dotprod(u,w)/vlength(u) = cos(1),
  dotprod(v,w)/vlength(v) = cos(1)};
```

$$eqs := \left\{ \frac{1}{14}(a + 2b + 3c)\sqrt{14} = \cos(1),\ \frac{1}{6}(2a + b + c)\sqrt{6} = \cos(1) \right\}$$

```
> soln := solve(eqs,{a,b});
```

$$soln := \left\{ a = \tfrac{1}{3}c + \tfrac{2}{3}\sqrt{6}\cos(1) - \tfrac{1}{3}\sqrt{14}\cos(1),\right.$$
$$\left. b = -\tfrac{1}{42}\sqrt{14}\left(5c\sqrt{14} + \sqrt{14}\sqrt{6}\cos(1) - 28\cos(1)\right) \right\}$$

```
> wc := subs(soln, evalm(w));
```

$wc :=$
$$\left[\tfrac{1}{3}c + \tfrac{2\sqrt{6}}{3}\cos(1) - \tfrac{\sqrt{14}}{3}\cos(1) \quad \tfrac{-\sqrt{14}}{42}\left(5c\sqrt{14} + \sqrt{14}\sqrt{6}\cos(1) - 28\cos(1)\right) \quad c\right]$$

Figure 7.2: | A vector making equal angles with two others. Part I |

```
> vlength(wc)^2 = 1;
```

$$\left(\tfrac{1}{3}c + \tfrac{2}{3}\sqrt{6}\cos{(1)} - \tfrac{1}{3}\sqrt{14}\cos{(1)}\right)^2 + \tfrac{1}{126}\left(5c\sqrt{14} + \sqrt{14}\sqrt{6}\cos{(1)} - 28\cos{(1)}\right)^2 + c^2$$
$$= 1$$

```
> c_values := solve(", c);
```

$$c\_values :=$$
$$\left\{c = -\tfrac{1}{5}\sqrt{6}\cos{(1)} + \tfrac{11}{35}\sqrt{14}\cos{(1)} + \tfrac{3}{35}\sqrt{7}\sqrt{-24\cos{(1)}^2 + 2\sqrt{6}\cos{(1)}^2\sqrt{14} + 5}\right\},$$
$$\left\{c = -\tfrac{1}{5}\sqrt{6}\cos{(1)} + \tfrac{11}{35}\sqrt{14}\cos{(1)} - \tfrac{3}{35}\sqrt{7}\sqrt{-24\cos{(1)}^2 + 2\sqrt{6}\cos{(1)}^2\sqrt{14} + 5}\right\}$$

```
> evalf(c_values);
```

$$\{c = .7854300975\}, \{c = -.0440796653\}$$

```
> subs(c_values[1], evalm(wc));
```

$$\left[\tfrac{3}{5}\sqrt{6}\cos{(1)} - \tfrac{8}{35}\sqrt{14}\cos{(1)} + \tfrac{1}{35}\%1 \right.$$
$$-\tfrac{1}{42}\sqrt{14}\left(5\left(-\tfrac{1}{5}\sqrt{6}\cos{(1)} + \tfrac{11}{35}\sqrt{14}\cos{(1)} + \tfrac{3}{35}\%1\right)\sqrt{14} + \sqrt{14}\sqrt{6}\cos{(1)} - 28\cos{(1)}\right)$$
$$\left. -\tfrac{1}{5}\sqrt{6}\cos{(1)} + \tfrac{11}{35}\sqrt{14}\cos{(1)} + \tfrac{3}{35}\%1\right]$$

$$\%1 := \sqrt{7}\sqrt{-24\cos{(1)}^2 + 2\sqrt{6}\cos{(1)}^2\sqrt{14} + 5}$$

```
> evalf(");
```

$$[.4702446317 \quad -.4024544048 \quad .7854300975]$$

```
> subs(c_values[2], evalm(wc));
```

$$\left[\tfrac{3}{5}\sqrt{6}\cos{(1)} - \tfrac{8}{35}\sqrt{14}\cos{(1)} + \tfrac{1}{35}\%1 \right.$$
$$-\tfrac{1}{42}\sqrt{14}\left(5\left(-\tfrac{1}{5}\sqrt{6}\cos{(1)} + \tfrac{11}{35}\sqrt{14}\cos{(1)} - \tfrac{3}{35}\%1\right)\sqrt{14} + \sqrt{14}\sqrt{6}\cos{(1)} - 28\cos{(1)}\right)$$
$$\left. -\tfrac{1}{5}\sqrt{6}\cos{(1)} + \tfrac{11}{35}\sqrt{14}\cos{(1)} - \tfrac{3}{35}\%1\right]$$

$$\%1 := \sqrt{7}\sqrt{-24\cos{(1)}^2 + 2\sqrt{6}\cos{(1)}^2\sqrt{14} + 5}$$

```
> evalf(");
```

$$[.1937413775 \quad .9800618669 \quad -.0440796653]$$

Figure 7.3: | A vector making equal angles with two others. Part II |

Now we must find the values of c that make $\mathbf{w}$ a unit vector. Squaring both sides of the equation $\sqrt{\mathbf{w} \cdot \mathbf{w}} = 1$ gives $\mathbf{w} \cdot \mathbf{w} = 1$, which will be easier to solve; enter `vlength(wc)^2 = 1` and then `solve`. We see there are two solutions in line 2 of Figure 7.3. To finish the problem, we approximate these numbers and substitute them back into `wc`. Line 4 is far too complicated to be useful. In line 4, Maple uses "%1" to replace a common subexpression, that is, to replace a portion of the expression that appears in several places, to improve readability. The `evalf` command returns an approximation for w in line 5.

Repeat this procedure (lines 6 and 7 of Figure 7.3) using the second solution for c to obtain the second vector and its approximation.

Notice that, geometrically, the set of all vectors that make an angle of 1 radian with a fixed vector forms a cone. Thus, we are looking for unit vectors that lie on the intersection of two cones with a common vertex.

Solved Problem 3: Find a vector $\mathbf{w}$ whose projections onto $\mathbf{u} := (1, 2, 3)$, $\mathbf{v} := (2, 1, 1)$, and $\mathbf{x} := (3, 1, 2)$ are $(2, 4, 6)$, $(4, 2, 2)$, and $(6, 2, 4)$, respectively.

Solution: First define each of the following vectors:

$$\mathbf{w} := (a, b, c) \quad \mathbf{u} := (1, 2, 3) \quad \mathbf{v} := (2, 1, 1) \quad \mathbf{x} := (3, 1, 2)$$

Notice that the projection of $\mathbf{w}$ onto $\mathbf{u}$ is $2\mathbf{u}$, the projection of $\mathbf{w}$ onto $\mathbf{v}$ is $2\mathbf{v}$, and the projection of $\mathbf{w}$ onto $\mathbf{x}$ is $2\mathbf{x}$. Thus we enter the matrix equation

$$[(\mathbf{w} \cdot \mathbf{u})/(\mathbf{u} \cdot \mathbf{u})\ \mathbf{u}\ \vdots\ (\mathbf{w} \cdot \mathbf{v})/(\mathbf{v} \cdot \mathbf{v})\ \mathbf{v}\ \vdots\ (\mathbf{w} \cdot \mathbf{x})/(\mathbf{x} \cdot \mathbf{x})\ \mathbf{x}] = [2\mathbf{u}\ \vdots\ 2\mathbf{v}\ \vdots\ 2\mathbf{x}]$$

Maple presents the matrix system of nine equations in three unknowns `eq` in line 2 of Figure 7.4. Now use `matsolve` to obtain the solution from line 3 of Figure 7.4, $\mathbf{w} = (3, -7, 13)$.

Solved Problem 4: Find all planes that pass through the points $(4, 1, 5)$ and $(2, 5, 3)$ and whose distance from the point $(1, 2, 4)$ is 2.

Solution: It is left to the reader to show that the required plane does not pass through the origin. (Hint: Find the equation of the plane through $(4, 1, 5)$, $(2, 5, 3)$, and $(0, 0, 0)$, and show that its distance from $(1, 2, 4)$ is not 2.) Thus, its equation can be written in the form $ax + by + cz + 1 = 0$. (Explain why.)

```
> w := vector([a,b,c]);
  u := vector([1,2,3]);
  v := vector([2,1,1]);
  x := vector([3,1,2]);
```

$$
\begin{aligned}
w &:= \begin{bmatrix} a & b & c \end{bmatrix} \\
u &:= \begin{bmatrix} 1 & 2 & 3 \end{bmatrix} \\
v &:= \begin{bmatrix} 2 & 1 & 1 \end{bmatrix} \\
x &:= \begin{bmatrix} 3 & 1 & 2 \end{bmatrix}
\end{aligned}
$$

```
> eq := augment(dotprod(w,u)/dotprod(u,u)*u,
  dotprod(w,v)/dotprod(v,v)*v,
  dotprod(w,x)/dotprod(x,x)*x)
  = augment(2*u,2*v,2*x);
```

$$
\begin{bmatrix}
\frac{1}{14}a + \frac{1}{7}b + \frac{3}{14}c & \frac{2}{3}a + \frac{1}{3}b + \frac{1}{3}c & \frac{9}{14}a + \frac{3}{14}b + \frac{3}{7}c \\
\frac{1}{7}a + \frac{2}{7}b + \frac{3}{7}c & \frac{1}{3}a + \frac{1}{6}b + \frac{1}{6}c & \frac{3}{14}a + \frac{1}{14}b + \frac{1}{7}c \\
\frac{3}{14}a + \frac{3}{7}b + \frac{9}{14}c & \frac{1}{3}a + \frac{1}{6}b + \frac{1}{6}c & \frac{3}{7}a + \frac{1}{7}b + \frac{2}{7}c
\end{bmatrix}
=
\begin{bmatrix}
2 & 4 & 6 \\
4 & 2 & 2 \\
6 & 2 & 4
\end{bmatrix}
$$

```
> matsolve(eq);
```

$$
\{c = 13, \ b = -7, \ a = 3\}
$$

Figure 7.4: | A vector with three given projections |

It is a fact that the distance from the point (x_0, y_0, z_0) to the plane $ax+by+cz+d = 0$ is $|ax_0 + by_0 + cz_0 + d| \, / \, |(a, b, c)|$. Therefore, the information given in the problem tells us the following.

$$
\begin{aligned}
4a + b + 5c + 1 &= 0 \\
2a + 5b + 3c + 1 &= 0 \\
\frac{|a + 2b + 4c + 1|}{|(a, b, c)|} &= 2
\end{aligned}
$$

The first step is to manipulate the third equation above by hand so as to make it easier for Maple to solve. Square both sides to eliminate the absolute values in the

```
>  eqs := {4*a+b+5*c+1=0, 2*a+5*b+3*c+1=0,
     (a+2*b+4*c)^2=4*(a^2+b^2+c^2) };
```

$$eqs := \{4a + b + 5c + 1 = 0, \ 2a + 5b + 3c + 1 = 0,$$
$$(a + 2b + 4c + 1)^2 = 4a^2 + 4b^2 + 4c^2\}$$

```
> solns := solve(eqs, {a,b,c});
```

$$solns := \left\{ b = -\tfrac{1}{9}\operatorname{RootOf}\left(283\ \_Z^2 - 46\ \_Z - 5\right) - \tfrac{1}{9}, \right.$$
$$\left. a = -\tfrac{11}{9}\operatorname{RootOf}\left(283\ \_Z^2 - 46\ \_Z - 5\right) - \tfrac{2}{9}, c = \operatorname{RootOf}\left(283\ \_Z^2 - 46\ \_Z - 5\right) \right\}$$

```
> allvalues(solns, 'd');
```

$$\left\{ a = -\tfrac{91}{283} - \tfrac{22}{283}\sqrt{6}, b = -\tfrac{34}{283} - \tfrac{2}{283}\sqrt{6}, c = \tfrac{23}{283} + \tfrac{18}{283}\sqrt{6} \right\},$$
$$\left\{ a = -\tfrac{91}{283} + \tfrac{22}{283}\sqrt{6}, b = -\tfrac{34}{283} + \tfrac{2}{283}\sqrt{6}, c = \tfrac{23}{283} - \tfrac{18}{283}\sqrt{6} \right\}$$

```
> evalf('');
```

$$\{c = .2370700190, \ b = -.1374522244, \ a = -.5119744676\},$$
$$\{c = -.07452584939, \ a = -.1311350730, \ b = -.1028304612\}$$

Figure 7.5: Planes through two points a distance 2 from a third point

numerator and the square root in the denominator, and then multiply to clear the fraction. Now enter and solve the system of equations

$$\left\{4a + b + 5c + 1 = 0, \ 2a + 5b + 3c + 1 = 0, \ (a + 2b + 4c + 1)^2 = 4\left(a^2 + b^2 + c^2\right)\right\}$$

using a, b, and c as *solve* variables. The solution is line 2 of Figure 7.5. Notice that the solution contains RootOf expressions; these result since there are several possible solutions. To find all solutions, use `allvalues` with the 'd' option. The function `allvalues` calculates the possible values, and the 'd' option keeps the value the same for a, b, and c. Finish by applying `evalf` to obtain numeric approximations. Thus, the required planes are

$$-.1311350730\ x - .1028304612\ y - .07452584939\ z + 1 = 0$$
$$-.5119744676\ x - .1374522244\ y + .2370700190\ z + 1 = 0$$

Solved Problem 5: Find all vectors of length 2 in $\mathbb{R}^4$ that are orthogonal to all three of $(1,5,2,8)$, $(3,5,7,9)$, and $(6,2,1,3)$.

<u>Solution</u>: First author each of the following as a separate expression:

$$u := [a, b, c, d] \quad x := [1, 5, 2, 8] \quad y := [3, 5, 7, 9] \quad z := [6, 2, 1, 3]$$

Just as in lower dimensions, vectors are orthogonal when their dot product is zero. Thus we enter and solve the system $\{u \cdot x = 0,\ u \cdot y = 0,\ u \cdot z = 0\}$. Using a, b, and c as *solve variables*, we obtain the solution as in line 3 of Figure 7.6. Substitute these values for a, b, and c into $\mathbf{u}$. Recall that the vectors we seek have length 2. To simplify the calculation, square both sides of $\sqrt{\mathbf{u} \cdot \mathbf{u}} = 2$. We enter and solve $\mathbf{u} \cdot \mathbf{u} = 4$. The solutions appear in line 5. Thus, we get the two vectors $\pm \frac{23}{\sqrt{445}} \left(\frac{1}{23}, -\frac{35}{23}, -\frac{5}{23}, 1 \right)$.

```
> u := vector([a,b,c,d]):
  x := vector([1,5,2,8]):
  y := vector([3,5,7,9]):
  z := vector([6,2,1,3]):
> eqs := {dotprod(u,x), dotprod(u,y), dotprod(u,z)};
```

$$eqs := \{a + 5b + 2c + 8d = 0,\ 3a + 5b + 7c + 9d = 0,\ 6a + 2b + c + 3d = 0\}$$

```
> abc := solve(eqs, {a,b,c});
```

$$abc := \left\{ b = -\frac{35}{23}d,\ c = -\frac{5}{23}d,\ a = \frac{1}{23}d \right\}$$

```
> u := subs(abc, evalm(u));
```

$$u := \left[\frac{1}{23}d \quad -\frac{35}{23}d \quad -\frac{5}{23}d \quad d \right]$$

```
> solve(dotprod(u,u)=4, {d});
```

$$\left\{ d = \frac{23}{445}\sqrt{445} \right\},\ \left\{ d = -\frac{23}{445}\sqrt{445} \right\}$$

Figure 7.6:   Vectors in $\mathbb{R}^4$ orthogonal to 3 given vectors

Solved Problem 6: For students who have studied calculus.

1. Establish the product rule for differentiation of dot products. That is,

$$\text{if } f(t) = [x(t), y(t)] \text{ and } g(t) = [z(t), w(t)],$$
$$\text{then } \tfrac{d}{dt}\{f(t) \cdot g(t)\} = f'(t) \cdot g(t) + f(t) \cdot g'(t)$$

2. Show how it follows quickly from 1 that $\frac{d}{dt}|f(t)|^2 = 2f(t) \cdot f'(t)$.

3. Show that the tip of the vector $\mathbf{v}$ is nearest the curve $g(t)$ at a point where the tangent vector to $g(t)$ is orthogonal to $\mathbf{v} - g(t)$.

4. Find the point on the curve $g(t) = (t, \frac{1}{t})$ that is nearest $(1, 0)$.

Solution: Parts 1—3 do not require a computer; however, see Exploration and Discovery Problem 2(h).

1.
$$\begin{aligned}
\tfrac{d}{dx}\{f(t) \cdot g(t)\} &= \tfrac{d}{dx}\{x(t)\,z(t) + y(t)\,w(t)\} \\
&= x'(t)\,z(t) + x(t)\,z'(t) + y'(t)\,w(t) + y(t)\,w'(t) \\
&= [x'(t), y'(t)] \cdot [z(t), w(t)] + [x(t), y(t)] \cdot [z'(t), w'(t)] \\
&= f'(t) \cdot g(t) + f(t) \cdot g'(t).
\end{aligned}$$

2. Observe that $|f(t)|^2 = f(t) \cdot f(t)$. Thus, setting $g(t) = f(t)$ in 1 above gives

$$\begin{aligned}
\tfrac{d}{dt}|f(t)|^2 &= \tfrac{d}{dt}\{f(t) \cdot f(t)\} \\
&= f'(t) \cdot f(t) + f(t) \cdot f'(t) \\
&= 2\,f(t) \cdot f'(t).
\end{aligned}$$

3. The distance from the tip of $\mathbf{v}$ to a point $g(t)$ on the curve is the length of the vector $\mathbf{v} - g(t)$. To minimize this distance, it is sufficient to minimize its square, $|\mathbf{v} - g(t)|^2$. This minimum occurs where the derivative is 0, so we have to solve $\frac{d}{dt}|\mathbf{v} - g(t)|^2 = 0$. According to 2 above, we can calculate the derivative as follows:

$$\frac{d}{dt}|\mathbf{v} - g(t)|^2 = 2(\mathbf{v} - g(t)) \cdot \frac{d}{dt}(\mathbf{v} - g(t)) = -2(\mathbf{v} - g(t)) \cdot g'(t)$$

Thus, the critical points occur where $(\mathbf{v} - g(t)) \cdot g'(t) = 0$, that is, where $(\mathbf{v} - g(t))$ is orthogonal to the tangent vector $g'(t)$.

```
>  g := t -> vector([t, 1/t]);
   dg := t -> vector([1, -1/t^2]);
```

$$g := t \rightarrow \text{vector}\left(\left[1, \tfrac{1}{t}\right]\right)$$

$$dg := t \rightarrow \text{vector}\left(\left[1, -\tfrac{1}{t^2}\right]\right)$$

```
>  y := dotprod((vector([1,0])-g(t)), dg(t));
```

$$y := 1 - t + \frac{1}{t^3}$$

```
>  plot(y, t=-5..5, -5..5);
```

```
>  fsolve(y, t=1..2);
```

$$1.380277569$$

```
>  g(");
```

$$[1.380277569, .7244919591]$$

Figure 7.7: A point on $\left(t, \tfrac{1}{t}\right)$ nearest $(1,0)$

4. We'll apply 3 above with $\mathbf{v} = [0,1]$. Define g and g's derivative, dg, with

$$g := t \rightarrow \text{vector}([t,1/t])$$
$$dg := t \rightarrow \text{vector}([1,-1/t^2])$$

If we simplify, we obtain line 3 of Figure 7.7, which must be solved. (Recall that solving an expression in Maple is the same as finding its zeros.) The exact solution is too complicated to use (try it), so we will solve it approximately. (Refer to Section A.7 of Appendix A for details.) Plot the graph of $\mathbf{y}$. From the image, it is clear that the point we seek is between $t = 1$ and $t = 2$. Ask Maple to solve on the interval $[1, 2]$ by using the *range option* for `fsolve`. The solution is displayed as line 4. Evaluate $g(1.38027)$ to obtain the answer displayed in line 6. As we can see from the graph, there is also a negative solution in the interval $[-1, 0]$. We leave it to the reader to find it and to interpret its meaning.

7.3   Exercises

1. Find all vectors of length 2 that make an angle of 1 radian with both $(1, 3, 1)$ and $(2, 1, 3)$.

2. Find a vector whose projections onto $(1, 2, 3)$, $(2, 1, 4)$, and $(3, 1, 1)$ are, respectively, $(3, 6, 9)$, $(-2, -1, -4)$, and $(6, 2, 2)$.

3. Find the equations of all planes that pass through the points $(3, 2, 1)$ and $(4, 1, 1)$ and are a distance 3 from $(6, 2, 7)$.

4. Let L denote the line through the origin and the point $(1, 2, 3)$. Find parametric equations of all lines that pass through the point $(7, 2, 1)$ and meet L at an angle of 1 radian.

5. Enter the vectors $x := [a, b, c]$ and $y := [r, s, t]$. Use Maple to verify that $\mathbf{x} \times \mathbf{y}$ is orthogonal to $\mathbf{x}$ and $\mathbf{y}$. Explain what you did. (Maple can be used to verify other facts and identities. See Exploration and Discovery Problem 2.)

6. Find all points on the intersection of the planes $3x + 7y - 9z = 8$ and $2x - 3y + 4z = 5$ that are a distance 6 from $(1, 2, 3)$.

7. Find all vectors of length 3 in $\mathbb{R}^5$ that are orthogonal to $(3, 6, 4, 2, 1)$, $(2, 6, 4, 2, 1)$, $(5, 1, 2, 7, 7)$, and $(9, 1, 2, 5, 7)$ simultaneously.

8. $\boxed{\textbf{For students who have studied calculus.}}$ Find the point on the curve

$$f := t \rightarrow [t, \ln(t)]$$

that is nearest the point $(1, 1)$.

7.4   Exploration and Discovery

1. Problems (a) and (b) below do not require a computer.

   (a) If (a, b, c), (d, e, f), and (g, h, i) are given points, show that

   $$\begin{vmatrix} 1 & x & y & z \\ 1 & a & b & c \\ 1 & d & e & f \\ 1 & g & h & i \end{vmatrix} = 0$$

   is the equation of a plane.

   (b) Show that the plane in part (a) passes through the points (a, b, c), (d, e, f), and (g, h, i). (You should be able to do this by substitution without expanding the determinant.)

   (c) Use this method to find the equation of the plane through the points $(1, 4, 9)$, $(8, 7, 2)$, and $(3, 3, 7)$.

   (d) Use this method to find the equation of the plane through the points $(1, 3, 7)$, $(2, 5, 9)$, and $(3, 8, 16)$. Explain what happens and why.

   (e) Show that $\begin{vmatrix} x & y & z \\ a & b & c \\ d & e & f \end{vmatrix} = 0$ is the equation of a plane that passes through
   the points (a, b, c) and (d, e, f). Prove that it must also pass through the origin.

2. Maple can be used to verify identities. Enter the vectors $\mathbf{x} := [a, b, c]$ and $\mathbf{y} := [r, s, t]$.

   (a) Expand $|\mathbf{x} - \mathbf{y}|^2 - |\mathbf{x}|^2 - |\mathbf{y}|^2$ in Maple. Now calculate $\mathbf{x} \cdot \mathbf{y}$. What relationship do you discover between $|\mathbf{x} - \mathbf{y}|^2 - |\mathbf{x}|^2 - |\mathbf{y}|^2$ and $\mathbf{x} \cdot \mathbf{y}$?

   (b) Enter and expand $|\mathbf{x} + \mathbf{y}|^2 + |\mathbf{x} - \mathbf{y}|^2$. How is it related to $|\mathbf{x}|$ and $|\mathbf{y}|$?

   (c) How is $\left(\dfrac{\mathbf{x} \cdot \mathbf{y}}{|\mathbf{x}| \, |\mathbf{y}|}\right)^2$ related to $\left(\dfrac{|\mathbf{x} \times \mathbf{y}|}{|\mathbf{x}| \, |\mathbf{y}|}\right)^2$?

   (d) Using $\mathbf{x} \cdot \mathbf{y} = |\mathbf{x}| \, |\mathbf{y}| \cos(\theta)$ and the preceding identity you discovered, express $|\mathbf{x} \times \mathbf{y}|$ in terms of $|\mathbf{x}|$, $|\mathbf{y}|$, and the angle θ. Be sure to explain how your answer follows from the previous identity (c).

(e) Try to discover other identities by this method.

(f) Enter the vectors $\mathbf{x} := [a, b]$ and $\mathbf{y} := [r, s]$. Try to verify identities (a), (b), and (c) you discovered above for $\mathbb{R}^2$. If anything strange happens, explain why.

(g) Enter the vectors $\mathbf{x} := [a, b, c, d]$ and $\mathbf{y} := [r, s, t, u]$. Try to verify identities (a), (b), and (c) that you discovered above for $\mathbb{R}^4$. If anything strange happens, explain why.

(h) In Solved Problem 6, we discussed several results involving calculus. It may seem a good idea to let Maple do them. Let's look at $\frac{d}{dt}|f(t)|^2 = 2\, f(t) \cdot f'(t)$. Enter the following four expressions:

```
z := vector([x(t), y(t)])
dz := map(diff, z, t)
2 * dotprod(z, dz) = diff(norm(z, 2)^2, t)
testeq('')
```

Does this match the result we had in 3 of Solved Problem 6? Explain the differences and why they occur.

Note that the expression $|x|\,/x$ is equal to

$$\text{signum}\,(x) := \left\{ \begin{array}{ll} 1 & \text{if } x > 0 \\ -1 & \text{if } x < 0 \\ \text{undefined} & \text{if } x = 0 \end{array} \right\}$$

7.5   Angles Between Vectors

LABORATORY EXERCISE 7–1

Name _____   Due Date _____

The <u>Problem</u>: Let $\mathbf{x} := [3, 2, 1]$ and $\mathbf{y} := [1, 2, 7]$. Follow the ideas in Solved Problem 1 to answer the following questions.

1. How many degrees are in an angle of one radian?

2. Find all *unit* vectors that make an angle of one radian with both $\mathbf{x}$ and $\mathbf{y}$.

3. Check your answers in (2) above.

4. Find all *unit* vectors that make an angle of one-half a radian with both $\mathbf{x}$ and $\mathbf{y}$. (*This problem has a pitfall. You should explain what difficulty you encounter here and why.*)

continued on next page

5. Describe geometrically the set of all vectors of the form $[1, 2, a]$.

6. Find all vectors of the form $[1, 2, a]$ (no restriction on the length) that make an angle of one radian with **x**.

Chapter 8
VECTOR SPACES AND SUBSPACES

LINEAR ALGEBRA CONCEPTS

- **Vector space**
- **Subspace**
- **Spaces of functions and matrices**
- **Linear combination**
- **Spanning set**
- **Nullspace**
- **Rank of a matrix**

8.1   Introduction

Many problems concerning general vector spaces can be reduced to questions about systems of linear equations, which Maple can help solve. We will use $\mathbb{R}^n$ to denote Euclidean n-space, $\mathbb{P}_n$ the space of polynomials of degree no greater than n, and $\mathbb{M}_{p,q}$ the space of p by q matrices.

Since Maple's **solve** does not handle matrix equations, we will again need the function **matsolve**. (See Appendix B.) After loading the **linalg** package, define the function:

```
matsolve := eq -> solve(convert(evalm(lhs(eq)-rhs(eq)), set))
```

8.2   Solved Problems

<u>Solved Problem 1</u>: Let S denote the subspace of $\mathbb{R}^4$ spanned by the vectors $\mathbf{u} = (1,2,3,4)$, $\mathbf{v} = (4,2,1,5)$ and $\mathbf{w} = (3,5,1,7)$. Determine if the vectors $\mathbf{x} = (8,9,5,16)$ and $\mathbf{y} = (7,2,1,3)$ are in S.

<u>Solution</u>: Enter and name the five vectors u := vector([1,2,3,4]), v := vector([4,2,1,5]), w := vector([3,5,1,7]), x := vector([8,9,5,16]) and y := vector([7,2,1,3]) as seen in Figure 8.1. We need to determine if **x** and **y** are linear combinations of the vectors **u**, **v**, and **w**.

Enter and evaluate as a matrix equation $\mathbf{x} = a\mathbf{u} + b\mathbf{v} + c\mathbf{w}$. If we ask Maple to solve this system of equations (remember to use `matsolve`, not `solve`) , we deduce from line 4 of Figure 8.1 that $\mathbf{x} = \mathbf{u} + \mathbf{v} + \mathbf{w}$. Thus, $\mathbf{x}$ is in S.

Repeat this procedure for the equation $\mathbf{y} = a\mathbf{u} + b\mathbf{v} + c\mathbf{w}$. We see from Maple's null response (line 7, Figure 8.1) that this equation has no solution. Thus, $\mathbf{y}$ is not in S.

```
>  u := vector([1,2,3,4]):
   v := vector([4,2,1,5]):
   w := vector([3,5,1,7]):
   x := vector([8,9,5,16]):
   y := vector([7,2,1,3]):
>  eq1 := x = a*u + b*v + c*w;
```

$$eq1 := x = au + bv + cw$$

```
>  evalm(eq1);
```

$$[8\ 9\ 5\ 16] = [a + 4b + 3c \quad 2a + 2b + 5c \quad 3a + b + c \quad 4a + 5b + 7c]$$

```
>  matsolve(eq1);
```

$$\{a = 1, b = 1, c = 1\}$$

```
>  eq2 := y = a*u + b*v + c*w;
```

$$eq2 := y = au + bv + cw$$

```
>  evalm(eq2);
```

$$[7\ 2\ 1\ 3] = [a + 4b + 3c \quad 2a + 2b + 5c \quad 3a + b + c \quad 4a + 5b + 7c]$$

```
>  matsolve(eq2);
>
```

Figure 8.1: The span of three vectors in $\mathbb{R}^4$

Alternative Solution: We can answer both questions by row-reducing the matrix in line 1 of Figure 8.2. This is an augmented matrix with the first three columns representing the coefficient matrix. The first four columns of the row-reduced form (line 3 of Figure 8.2) give us the solution we got in line 4 of Figure 8.1. The first three columns along with the last one in the row-reduced form show that there is no solution in the second case. Be sure you understand why this alternative solution works.

```
> A := augment(u,v,w,x,y);
```

$$A := \begin{bmatrix} 1 & 4 & 3 & 8 & 7 \\ 2 & 2 & 5 & 9 & 2 \\ 3 & 1 & 1 & 5 & 1 \\ 4 & 5 & 7 & 16 & 3 \end{bmatrix}$$

```
> rref(A);
```

$$\begin{bmatrix} 1 & 0 & 0 & 1 & 0 \\ 0 & 1 & 0 & 1 & 0 \\ 0 & 0 & 1 & 1 & 0 \\ 0 & 0 & 0 & 0 & 1 \end{bmatrix}$$

Figure 8.2: The span of three vectors in $\mathbb{R}^4$: Alternative solution

Solved Problem 2: Find a vector in $\mathbb{R}^5$ that is not in the subspace spanned by the vectors $(1, 2, 3, 2, 1)$, $(3, 2, 7, 1, 6)$, $(2, 5, 6, 1, 4)$, and $(4, 1, 7, 2, 2)$.

Solution: We seek a vector $\mathbf{v} = (p, q, r, s, t)$ so that the following equation has no solution.

$$a\ (1, 2, 3, 2, 1) + b\ (3, 2, 7, 1, 6) + c\ (2, 5, 6, 1, 4) + d\ (4, 1, 7, 2, 2) = (p, q, r, s, t)$$

Enter $\mathbf{v}$ and the four vectors as v1 through v4, and consider the expression mateq

```
a*u1 + b*u2 + c*u3 + d*u4 - v
```

as seen in lines 1 and 2 of Figure 8.3. Before we ask Maple to solve, mateq must be converted to a set. Since there are five equations in nine unknowns, we must select the solve variables. In Figure 8.3 we have chosen solve variables, a, b, c, d, and p. From line 4 we see $p = -\frac{1}{4}\ s - \frac{1}{4}\ q + \frac{3}{4}\ r - \frac{1}{4}\ t$. Thus, any vector $\mathbf{v}$ that does *not* satisfy this equation will do. We use $\mathbf{v} = (0, 0, 0, 0, 1)$ as an example.

```
>  v := vector([p,q,r,s,t]):
   u1 := vector([1,2,3,2,1]):
   u2 := vector([3,2,7,1,6]):
   u3 := vector([2,5,6,1,4]):
   u4 := vector([4,1,7,2,2]):
> mateq := evalm( a*u1 + b*u2 + c*u3 + d*u4 - v );
```

$$mateq := [a + 3b + 2c + 4d - p \quad 2a + 2b + 5c + d - q \quad 3a + 7b + 6c + 7d - r$$
$$2a + b + c + 2d - s \quad a + 6b + 4c + 2d - t]$$

```
> sys := convert(mateq, set);
```

$$sys := \{a + 3b + 2c + 4d - p \quad 2a + 2b + 5c + d - q \quad 3a + 7b + 6c + 7d - r$$
$$2a + b + c + 2d - s \quad a + 6b + 4c + 2d - t\}$$

```
> solve(sys, {a,b,c,d,p});
```

$$\left\{ d = -\frac{35}{124}\,t - \frac{27}{124}\,s - \frac{11}{124}\,q + \frac{37}{124}\,r, \quad c = -\frac{19}{124}\,t - \frac{43}{124}\,s + \frac{33}{124}\,q - \frac{19}{124}\,r \right.$$
$$p = -\frac{1}{4}\,s - \frac{1}{4}\,q + \frac{3}{4}\,r - \frac{1}{4}\,t, \quad a = \frac{6}{31}\,t + \frac{25}{31}\,s + \frac{1}{31}\,q - \frac{9}{31}\,r$$
$$\left. b = \frac{41}{124}\,t + \frac{21}{124}\,s - \frac{19}{124}\,q - \frac{15}{124}\,r \right\}$$

Figure 8.3: A vector not in a given subspace

Alternative solution using Gaussian Elimination: The augmented matrix of the system of equations is

$$A = \begin{bmatrix} 1 & 3 & 2 & 4 & p \\ 2 & 2 & 5 & 1 & q \\ 3 & 7 & 6 & 7 & r \\ 2 & 1 & 1 & 2 & s \\ 1 & 6 & 4 & 2 & t \end{bmatrix}$$

We continue from Figure 8.3 by defining the 5 by 5 matrix A in line 1 of Figure 8.4 with augment. Apply gausselim to get the result in line 2 of Figure 8.4. From the last row of the matrix, we see that the system of equations has a solution if and only if $t + 4p + q - 3r + s = 0$, and we arrive at the previous solution.

```
> A := augment(u1,u2,u3,u4,v);
```

$$A := \begin{bmatrix} 1 & 3 & 2 & 4 & p \\ 2 & 2 & 5 & 1 & q \\ 3 & 7 & 6 & 7 & r \\ 2 & 1 & 1 & 2 & s \\ 1 & 6 & 4 & 2 & t \end{bmatrix}$$

```
> gausselim(A);
```

$$\begin{bmatrix} 1 & 3 & 2 & 4 & p \\ 0 & -4 & 1 & -7 & q - 2p \\ 0 & 0 & \frac{-1}{2} & \frac{-3}{2} & r - 2p - \frac{1}{2}q \\ 0 & 0 & 0 & \frac{31}{2} & s + \frac{35}{2}p + 3q - \frac{17}{2}r \\ 0 & 0 & 0 & 0 & t + 4p + q - 3r + s \end{bmatrix}$$

Figure 8.4: Alternative solution to Solved Problem 2

Solved Problem 3: Determine if the five vectors below span $\mathbb{R}^4$.

$$(2,1,3,5),\ (4,3,7,2),\ (3,1,2,5),\ (2,1,5,6),\ (4,1,3,1)$$

Solution: These five vectors span $\mathbb{R}^4$ if and only if the following equation has a solution for each vector **v** in $\mathbb{R}^4$.

$$a\ (2,1,3,5) + b\ (4,3,7,2) + c\ (3,1,2,5) + d\ (2,1,5,6) + e\ (4,1,3,1) = \mathbf{v}$$

Enter as in Figure 8.5. As we've suggested before, you may want to define each vector on a separate line and refer to them by name in the equation rather than type the whole thing.

Ask Maple to solve this problem, using the same procedure as before, selecting the *solve variables* a, b, c, d, and e. We see from line 4 of Figure 8.5 that the solution does not place any restrictions on p, q, r, or s. We conclude that there is a solution no matter what their values are, and so the five given vectors span $\mathbb{R}^4$.

```
> v := vector([p,q,r,s]):    u1 := vector([2,1,3,5]):
  u2 := vector([4,3,7,2]):   u3 := vector([3,1,2,5]):
  u4 := vector([2,1,5,6]):   u5 := vector([4,1,3,1]):
> mateq := evalm(a*u1 + b*u2 + c*u3 + d*u4 + e*u5 - v);
```

$$mateq := [2a + 4b + 3c + 2d + 4e - p \quad a + 3b + c + d + e - q$$
$$3a + 7b + 2c + 5d + 3e - r \quad 5a + 2b + 5c + 6e + d - s]$$

```
> sys := convert(mateq, set);
```

$$sys := \{2a + 4b + 3c + 2d + 4e - p \quad a + 3b + c + d + e - q$$
$$3a + 7b + 2c + 5d + 3e - r \quad 5a + 2b + 5c + 6e + d - s\}$$

```
> solve(sys, {a,b,c,d,e});
```

$$\left\{ e = -\tfrac{11}{5}b + \tfrac{1}{2}q + \tfrac{1}{10}p - \tfrac{1}{5}s + \tfrac{1}{10}r, \quad d = \tfrac{21}{5}b - 3q + \tfrac{2}{5}p + \tfrac{1}{5}s + \tfrac{2}{5}r, \right.$$
$$\left. c = \tfrac{32}{5}b - 3q + \tfrac{4}{5}p + \tfrac{2}{5}s - \tfrac{1}{5}r, \quad a = -\tfrac{57}{5}b + \tfrac{13}{2}q - \tfrac{13}{10}p - \tfrac{2}{5s} - \tfrac{3}{10}r, \quad b = b \right\}$$

Figure 8.5: | A spanning set for $\mathbb{R}^4$ |

Alternative solution using the *rank* of a matrix: The vector equation above is equivalent to the following matrix equation.

$$\begin{bmatrix} 2 & 4 & 3 & 2 & 4 \\ 1 & 3 & 1 & 1 & 1 \\ 3 & 7 & 2 & 5 & 3 \\ 5 & 2 & 5 & 6 & 1 \end{bmatrix} \begin{bmatrix} a \\ b \\ c \\ d \\ e \end{bmatrix} = \begin{bmatrix} p \\ q \\ r \\ s \end{bmatrix}$$

Continuing from Figure 8.5, use **augment** to obtain the coefficient matrix A (see line 1 of Figure 8.6) and then row-reduce it. The row echelon form is in line 2 of Figure 8.6. Notice that there is a "1" in each row of the reduced form of the coefficient matrix. This ensures that the system has a solution. We conclude once more that the five given vectors span $\mathbb{R}^4$.

```
> A := augment(u1,u2,u3,u4);
```

$$A := \begin{bmatrix} 2 & 4 & 3 & 2 & 4 \\ 1 & 3 & 1 & 1 & 1 \\ 3 & 7 & 2 & 5 & 3 \\ 5 & 2 & 5 & 6 & 1 \end{bmatrix}$$

```
> rref(A);
```

$$\begin{bmatrix} 1 & 0 & 0 & 0 & \frac{-57}{11} \\ 0 & 1 & 0 & 0 & \frac{5}{11} \\ 0 & 0 & 1 & 0 & \frac{32}{11} \\ 0 & 0 & 0 & 1 & \frac{21}{11} \end{bmatrix}$$

Figure 8.6: $\boxed{\text{Alternative method of showing that vectors span } \mathbb{R}^4}$

The crucial observation in the work above is that each row of the reduced form of the coefficient matrix contains a leading 1. This leads to the following:

Definition: The *rank* of a matrix A is the number of leading 1's (or the number of nonzero rows) in the reduced echelon form of A.

The rank is easily calculated using Maple. Simply row-reduce and count the number of nonzero rows. (Also see ?rank.) The above observation yields the following at once:

Theorem: Let A be a matrix with n rows. The columns of A span $\mathbb{R}^n$ if and only if the rank of A is n.

$\boxed{\text{To determine if a set of vectors spans } \mathbb{R}^n \text{, arrange them as columns of a matrix. The vectors span } \mathbb{R}^n \text{ if and only if the rank of this matrix is } n.}$

Solved Problem 4: Let S denote the subspace of $M_{2,2}$ consisting of matrices of the form $\begin{bmatrix} a+b+c & a-c \\ 3b+3c & 2a-b+4c \end{bmatrix}$, and let T denote the subspace of $M_{2,2}$ consisting of matrices of the form $\begin{bmatrix} p-q+3r & p+2q \\ r-q & p-q-r \end{bmatrix}$. Describe all matrices that are in both S and T.

Solution: Enter and name the 2 by 2 matrices above as in line 1 of Figure 8.7. We wish to find choices of a, b, c and p, q, r that make these two matrices the same. Instead of working with $u = v$, we will use $u - v$ as in line 2 in Figure 8.6. Change the matrix expression to a set with **convert**. Next solve the system of equations in line 4 of Figure 8.6 using solve variables a, b, c, and p. We see that the original equation has a solution if and only if $p = \frac{40}{3}q - \frac{58}{3}r$. Substitute this value for p into **v**. Those vectors that are in both subspaces are of the form displayed in line 5 of Figure 8.6. The matrices we have described form the intersection of two subspaces, which should again be a subspace. The description shows this.

```
> u := matrix(2,2,[a+b+c, a-c, 3*b+3*c, 2*a-b+4c]);
  v := matrix(2,2,[p-q+3*r, p+2*q, r-q, p-q-r]);
```

$$u := \begin{bmatrix} a+b+c & a-c \\ 3b+3c & 2a-b+4c \end{bmatrix}$$

$$v := \begin{bmatrix} p-q+3r & P+2q \\ r-q & p-q-r \end{bmatrix}$$

```
> matsys := evalm(u-v);
```

$$\begin{bmatrix} a+b+c-p+q-3r & a-c-p-2q \\ 3b+3c-r+q & 2a-b+4c-p+q+r \end{bmatrix}$$

```
> sys := convert(matsys, set);
```

$sys := \{a+b+c-p+q-3r,\ a-c-p-2q,\ 3b+3c-r+q,\ 2a-b+4c-p+q+r\}$

```
> solve(sys, {a,b,c,p});
```

$$\left\{ p = \frac{40}{3}q - \frac{58}{3}r,\ a = \frac{38}{3}q - \frac{50}{3}r,\ b = \frac{7}{3}q - \frac{7}{3}r,\ c = -\frac{8}{3}q + \frac{8}{3}r \right\}$$

```
> subs(p=40/3*q-58/3*r, evalm(v));
```

$$\begin{bmatrix} \frac{37}{3}q - \frac{49}{3}r & \frac{46}{3}q - \frac{58}{3}r \\ r-q & \frac{37}{3}q - \frac{61}{3}r \end{bmatrix}$$

Figure 8.7: | The intersection of subspaces of $\mathbb{M}_{2,2}$ |

Solved Problem 5: Let $p := x^3 + x^2 + 1$, $q := 2x^3 - x^2 - 3x + 1$, and $r := x^3 - x + 4$, and let S denote the subspace of $\mathbb{P}_3$ spanned by p, q, and r. For what values of a, b, c, and d does $ax^3 + bx^2 + cx + d$ belong to S?

Solution: First define the polynomials with the expressions $p := x^3 + x^2 + 1$, $q := 2x^3 - x^2 - 3x + 1$, $r := x^3 - x + 4$, and $v := ax^3 + bx^2 + cx + d$ as done in line 1 of Figure 8.8. We wish to know what restrictions must be placed on a, b, c, and d to ensure that we can solve $ax^3 + bx^2 + cx + d = e\mathbf{p} + f\mathbf{q} + g\mathbf{r}$ for e, f, and g. To simplify the typing, we will work with the expression $e\mathbf{p} + f\mathbf{q} + g\mathbf{r} - \mathbf{v}$. To collect like terms with respect to the variable x, use Maple's `collect` function. The result of collecting like terms appears in line 3 of Figure 8.8. Applying the `coeffs` function returns the coefficients of the powers of x in sequence. We will put the coefficients into a set in line 4 of Figure 8.8. Examining the coefficients yields the following system of equations:

```
> p := x^3 + x^2 + 1:
  q := 2*x^3 - x^2 - 3*x + 1:
  r := x^3 - x + 4:
  v := a*x^3 + b*x^2 + c*x + d:
> eq := e*p + f*q + g*r - v;
```

$$eq := e\left(x^3 + x^2 + 1\right) + f\left(2x^3 - x^2 - 3x + 1\right) + g\left(x^3 - x + 4\right) - ax^3 - bx^2 - cx - d$$

```
> collect(eq, x);
```

$$(e + 2f - a + g)\,x^3 + (e - f - b)\,x^2 + (-3f - g - c)\,x + e - d + f + 4g$$

```
> sys := {coeffs('', x)};
```

$$sys := \{e + 2f - a + g,\ e - f - b,\ -3f - g - c,\ e - d + f + 4g\}$$

```
> solve(sys, {e,f,g});
> solve(sys, {e,f,g,a});
```

$$\left\{e = -\frac{1}{10}d + \frac{11}{10}b - \frac{2}{5}c,\ f = -\frac{1}{10}d + \frac{1}{10}b - \frac{2}{5}c,\ g = \frac{3}{10}d - \frac{3}{10}b + \frac{1}{5}c,\ a = b - c\right\}$$

Figure 8.8: $\boxed{\text{A subspace of } \mathbb{P}_3}$

$$a = e + 2f + g$$
$$b = e - f$$
$$c = -3f - g$$
$$d = e + f + 4g$$

The null result from `solve` in line 5 indicates that Maple cannot solve this system of equations using only the *solve variables* e, f, and g. (This indicates that some restrictions must be placed on a, b, c, and d to guarantee a solution.) The solution using *solve variables* e, f, g, and a appears in line 6 of Figure 8.8. We conclude from line 6 that a solution exists provided $a = b - c$. Thus, polynomials in S are of the form $(b - c)x^3 + bx^2 + cx + d$.

Solved Problem 6: Find a spanning set for the nullspace of the matrix

$$A = \begin{bmatrix} 2 & 7 & 8 & 4 & 1 \\ 3 & 2 & 5 & 1 & 9 \\ 4 & 1 & 3 & 8 & 1 \end{bmatrix}$$

Solution: The nullspace of A is the collection of all solutions of the equation $A\mathbf{x} = \mathbf{0}$. Enter and name the matrix A. To facilitate back-substituting later, row-reduce the augmented matrix $[A|\,\mathbf{0}]$. Using `backsub` in line 3 of Figure 8.9, we see that the solution is

$$x_1 = -\frac{11}{3}t_4 + \frac{97}{39}t_5, \quad x_2 = -\frac{10}{3}t_4 + \frac{209}{39}t_5, \quad x_3 = \frac{10}{3}t_4 - \frac{212}{39}t_5, \quad x_4 = t_4, \quad x_5 = t_5$$

Thus,

$$\mathbf{x} = \begin{bmatrix} -\frac{11}{3}x_4 + \frac{97}{39}x_5 \\ -\frac{10}{3}x_4 + \frac{209}{39}x_5 \\ \frac{10}{3}x_4 - \frac{212}{39}x_5 \\ x_4 \\ x_5 \end{bmatrix} = x_4 \begin{bmatrix} \frac{-11}{3} \\ \frac{-10}{3} \\ \frac{10}{3} \\ 1 \\ 0 \end{bmatrix} + x_5 \begin{bmatrix} \frac{97}{39} \\ \frac{209}{39} \\ \frac{-212}{39} \\ 0 \\ 1 \end{bmatrix}$$

This says that each vector in the nullspace of A is a linear combination of the vectors

$$\begin{bmatrix} \frac{-11}{3} \\ \frac{-10}{3} \\ \frac{10}{3} \\ 1 \\ 0 \end{bmatrix} \quad \text{and} \quad \begin{bmatrix} \frac{97}{39} \\ \frac{209}{39} \\ \frac{-212}{39} \\ 0 \\ 1 \end{bmatrix}$$

Notice that the first vector is obtained by setting $x_4 = 1$ and $x_5 = 0$ and that the second is obtained by setting $x_4 = 0$ and $x_5 = 1$ (see Figure 8.9). Thus, these two vectors belong to the nullspace of A and form a spanning set. (Note: It is wise to check your work by multiplying the two vectors by A to see if you do indeed get the zero vector. We leave this to the reader.) This exercise indicates a procedure for finding a spanning set for the nullspace of a matrix.

**Procedure for Constructing a Spanning Set
for the Nullspace of a Matrix**

- Row-reduce the matrix to obtain the solution of $A\mathbf{x} = \mathbf{0}$.
- Obtain a solution corresponding to each independent variable by setting that variable to 1 and the other independent variables to 0.
- Collect the vectors obtained to form a spanning set for the nullspace.

```
> A := matrix(3,5, [2,7,8,4,1, 3,2,5,1,9, 4,1,3,8,1]);
```

$$A := \begin{bmatrix} 2 & 7 & 8 & 4 & 1 \\ 3 & 2 & 5 & 1 & 9 \\ 4 & 1 & 3 & 8 & 1 \end{bmatrix}$$

```
> rref(augment(A, vector([0,0,0])));
```

$$\begin{bmatrix} 1 & 0 & 0 & \frac{11}{3} & -\frac{97}{39} & 0 \\ 0 & 1 & 0 & \frac{10}{3} & -\frac{209}{39} & 0 \\ 0 & 0 & 1 & -\frac{10}{3} & \frac{212}{39} & 0 \end{bmatrix}$$

```
> ns := backsub(");
```

$$ns := \left[-\frac{11}{34}t4 + \frac{97}{35}t5 \quad -\frac{10}{3}t4 + \frac{209}{39}t5 \quad t4 \quad t5 \right]$$

```
> subs(t4=1, t5=0, evalm(ns));
```

$$\left[\frac{-11}{3} \quad \frac{-10}{3} \quad \frac{10}{3} \quad 1 \quad 0 \right]$$

```
> subs(t4=0, t5=1, evalm(ns));
```

$$\left[\frac{97}{39} \quad \frac{209}{39} \quad \frac{-212}{39} \quad 0 \quad 1 \right]$$

Figure 8.9: A spanning set for the nullspace of a matrix

8.3 Exercises

1. Let S denote the subspace of $\mathbb{R}^5$ spanned by $(2, 3, 1, 6, 4)$, $(3, 5, 1, 2, 3)$, $(4, 2, 1, 5, 7)$, and $(3, 8, 2, 2, 1)$.

   (a) Determine if $(6, 2, 1, 11, 13)$ is in S.

   (b) For what values of a, b, c, d, and e does (a, b, c, d, e) belong to S?

   (c) Find the intersection of S with the subspace of $\mathbb{R}^5$ spanned by $(2, 9, 1, 8, 8)$, $(4, 1, 6, 7, 7)$, and $(3, 5, 7, 9, 1)$.

   (d) Find a vector in $\mathbb{R}^5$ that is not in S.

2. Let U denote the subspace of $\mathbb{P}_4$ spanned by the polynomials $p := x^4 - 3x^3 + x^2 - x + 1$, $r := 2x^4 - 3x^2 + x + 3$, $s := x^4 + x^3 + x^2 + 4x - 1$, and $t := 3x^4 + x^3 - 2x^2 + 3$.

   (a) Determine if $x^4 - 3x^3 + 4x^2 + 2x - 7$ is in U.

   (b) For what values of a, b, c, d, and e does $ax^4 + bx^3 + cx^2 + dx + e$ belong to U?

3. Let T be the subspace of $\mathbb{M}_{3,3}$ consisting of matrices of the form

$$\begin{bmatrix} a & b & c \\ a - b & a + c & b - 2c \\ a + b + c & a + 3c & 2a + b \end{bmatrix}.$$

Find the intersection of T with the subspace of $\mathbb{M}_{3,3}$ consisting of matrices of the form $\begin{bmatrix} a & b & a + b \\ c & a + c & d \\ e & c + d + e & f \end{bmatrix}$.

4. $\boxed{\text{This exercise may be solved using rank.}}$ Determine if the following sets of vectors span $\mathbb{R}^5$.

   (a) $(1, 3, 2, 5, 6)$, $(3, 3, 2, 1, 5)$, $(5, 6, 8, 1, 2)$, $(3, 6, 5, 1, 8)$, $(5, 5, 2, 6, 7)$

   (b) $(1, 3, 2, 1, 1)$, $(2, 1, 1, 3, 1)$, $(1, 1, 2, 2, 3)$, $(3, 4, 3, 4, 2)$, $(3, 2, 3, 5, 4)$

5. | **This exercise may be solved using rank.** | Determine if the matrices below span $\mathbb{M}_{2,2}$.

$$\begin{bmatrix} 3 & 7 \\ 2 & 5 \end{bmatrix}, \quad \begin{bmatrix} 5 & 9 \\ 2 & 3 \end{bmatrix}, \quad \begin{bmatrix} 6 & 6 \\ 1 & 2 \end{bmatrix}, \quad \begin{bmatrix} 4 & 5 \\ 9 & 1 \end{bmatrix}$$

6. Find spanning sets for the nullspaces of the following matrices.

(a) $\begin{bmatrix} 3 & 5 & 9 & 8 \\ 2 & 3 & 1 & 7 \\ 4 & 1 & 2 & 9 \\ 6 & 3 & 4 & 1 \end{bmatrix}$ (b) $\begin{bmatrix} 2 & 3 & 4 & 1 & 2 \\ 5 & 2 & 4 & 8 & 1 \\ 3 & 1 & 2 & 5 & 4 \\ 8 & 3 & 6 & 13 & 5 \end{bmatrix}$

7. If $f(x)$ is a function and r is a number, a new function can be defined by $f(x-r)$. It is called a *translate* of f. Let $f := x \rightarrow x^2$. If r, s, and t are distinct real numbers, show that $\{f(x-r), f(x-s), f(x-t)\}$ is a spanning set for $\mathbb{P}_2$.

8.4 Exploration and Discovery

1. **For students who have studied calculus.** Let A be the matrix given by

$$A := \begin{bmatrix} 0 & 3 & 0 & 0 \\ 0 & 0 & 2 & 0 \\ 0 & 0 & 0 & 1 \\ 0 & 0 & 0 & 0 \end{bmatrix}$$

(a) Find the nullspace of A.

(b) Find the rank of A.

(c) Define the vectors $\mathbf{v} = (3, 2, -1, 4)$ and $\mathbf{p} = (x^3, x^2, x, 1)$. Compare $\mathbf{v} \cdot \mathbf{p}$ with $(A\mathbf{v}) \cdot \mathbf{p}$. What do you observe?

(d) Take several randomly chosen vectors $\mathbf{u}$ and compare $\mathbf{u} \cdot \mathbf{p}$ with $(A\mathbf{u}) \cdot \mathbf{p}$. What standard calculus operation does A perform?

(e) Give an interpretation of the nullspace of A in light of your answer to d.

8.5 Describing Sets of Vectors

LABORATORY EXERCISE 8–1

Name _____ Due Date _____

Let $A = \begin{bmatrix} 3 & 5 & 9 & 8 \\ 2 & 3 & 1 & 7 \\ 4 & 1 & 2 & 9 \\ 6 & 3 & 4 & 1 \end{bmatrix}$. Let S denote the subspace of $\mathbb{R}^4$ spanned by the vectors $(3,2,1,1)$, $(1,1,3,1)$, $(2,1,-2,0)$, and $(4,3,4,2)$.

1. Find a spanning set for the nullspace of A.

2. Find all vectors that are in both S and the nullspace of A.

Chapter 9
INDEPENDENCE, BASIS, AND DIMENSION

LINEAR ALGEBRA CONCEPTS

- **Linearly independent set**
- **Basis**
- **Dimension**
- **Coordinate vector**

9.1 Introduction

A vector space has many spanning sets. A spanning set that is "minimal" in the sense that it does not contain unnecessary vectors is a *basis*. The number of vectors in a basis is called the *dimension* of the vector space.

An element $\mathbf{v}$ in a vector space with a basis S can be written as a linear combination of the elements of S in a unique way, provided that we acknowledge that the elements of S are in a prescribed order. When the coefficients in this linear combination are written as a vector, it is called the coordinate vector of $\mathbf{v}$ relative to the *ordered* basis S. When we discuss the coordinate vector of $\mathbf{v}$ relative to a basis S, we will always assume S is ordered even though ordinary set notation is used.

For example, the coordinate vector of the polynomial $\mathbf{v} = x^2 - x$ relative to the ordered basis $\{x^2, x, 1\}$ is $(1, -1, 0)$, and the coordinate vector of $x^2 - x$ relative to the ordered basis $\{x, 1, x^2\}$ is $(-1, 0, 1)$.

We will need the function `matsolve` from Appendix B. After loading the `linalg` package, define:

```
matsolve := eq -> solve(convert(evalm(lhs(eq)-rhs(eq)), set))
```

9.2 Solved Problems

Solved Problem 1: Determine if the following vectors are linearly independent:

$$(1, 4, 6, 8, 7, 2), \ (5, 2, 9, 1, 3, 5), \ (3, 6, 8, 1, 4, 7), \ (5, 5, 2, 8, 6, 3)$$

<u>Solution</u>: Enter the following vectors:

$$u := [1, 4, 6, 8, 7, 2], \quad v := [5, 2, 9, 1, 3, 5], \quad w := [3, 6, 8, 1, 4, 7], \quad x := [5, 5, 2, 8, 6, 3]$$

(See Figure 9.1.) We want to determine whether there is a unique solution to $au + bv + cw + dx = 0$. Recall that if Maple is asked to solve an expression that is not an equation, it sets the expression equal to zero before solving. However, our function `matsolve` expects an equation. Thus, you must enter the equation `a*u + b*v + c*w + d*x = 0` as in line 2.

<u>Warning</u>: It is common practice to write $au + bv + cw + dx = 0$ and assume the reader will know from the context that 0 must be the zero vector in $\mathbb{R}^6$ and not the number 0. It's the only way the statement makes sense.

```
> u := vector([1,4,6,8,7,2]);
  v := vector([5,2,9,1,3,5]);
  w := vector([3,6,8,1,4,7]);
  x := vector([5,5,2,8,6,3]);
```

$$
\begin{aligned}
u &:= [1, 4, 6, 8, 7, 2] \\
v &:= [5, 2, 9, 1, 3, 5] \\
w &:= [3, 6, 8, 1, 4, 7] \\
x &:= [5, 5, 2, 8, 6, 3]
\end{aligned}
$$

```
> eq := evalm(a*u + b*v + c*w + d*x) = 0;
```

$$eq := [a + 5b + 3c + 5d \quad 4a + 2b + 6c + 5d \quad 6a + 9b + 8c + 2d \\ 8a + b + c + 8d \quad 7a + 3b + 4c + 6d \quad 2a + 5b + 7c + 3d] = 0$$

```
> matsolve(eq);
```

$$\{a = 0, \quad b = 0, \quad c = 0, \quad d = 0\}$$

Figure 9.1: $\boxed{\text{Linearly independent vectors in } \mathbb{R}^6}$

<u>Alternative Solution</u>: Another method for determining if vectors in $\mathbb{R}^n$ are linearly independent is to make a matrix using the given vectors as the columns. Of course, this is just the coefficient matrix of the system we solved previously.

> *The vectors are linearly independent if and only if the rank of this matrix is the same as the number of vectors.*

Taking advantage of Maple's **augment**, enter the vectors as the 6 by 4 matrix A as done in line 1 of Figure 9.2.

$$A := \begin{bmatrix} 1 & 5 & 3 & 5 \\ 4 & 2 & 6 & 5 \\ 6 & 9 & 8 & 2 \\ 8 & 1 & 1 & 8 \\ 7 & 3 & 4 & 6 \\ 2 & 5 & 7 & 3 \end{bmatrix}$$

Now ask Maple to row-reduce. From line 2 of Figure 9.2 we see that the rank of the matrix is the same as the number of vectors (4), and we conclude once more that the vectors are linearly independent.

```
> A := augment([1,4,6,8,7,2],[5,2,9,1,3,5],[3,6,8,1,4,7],
      [5,5,2,8,6,3]);
```

$$A := \begin{bmatrix} 1 & 5 & 3 & 5 \\ 4 & 2 & 6 & 5 \\ 6 & 9 & 8 & 2 \\ 8 & 1 & 1 & 8 \\ 7 & 3 & 4 & 6 \\ 2 & 5 & 7 & 3 \end{bmatrix}$$

```
> rref(A);
```

$$\begin{bmatrix} 1 & 0 & 0 & 0 \\ 0 & 1 & 0 & 0 \\ 0 & 0 & 1 & 0 \\ 0 & 0 & 0 & 1 \\ 0 & 0 & 0 & 0 \\ 0 & 0 & 0 & 0 \end{bmatrix}$$

Figure 9.2: Test for independence using rank

Solved Problem 2: Determine if the following elements of $\mathbb{M}_{4,2}$ are linearly independent. If they are linearly dependent, write one as a linear combination of the others.

$$\mathbf{U} := \begin{bmatrix} 2 & 3 \\ 1 & 4 \\ 3 & 2 \\ 1 & 5 \end{bmatrix} \quad \mathbf{V} := \begin{bmatrix} 1 & 5 \\ 4 & 2 \\ 3 & 1 \\ 2 & 2 \end{bmatrix} \quad \mathbf{X} := \begin{bmatrix} 3 & 8 \\ 5 & 6 \\ 6 & 3 \\ 3 & 7 \end{bmatrix} \quad \mathbf{Y} := \begin{bmatrix} 1 & -2 \\ -3 & 2 \\ 0 & 1 \\ -1 & 3 \end{bmatrix}$$

Solution: Define each of the matrices. We want to set $a\mathbf{U} + b\mathbf{V} + c\mathbf{X} + d\mathbf{Y}$ equal to the zero matrix and solve for a,b,c, and d. We see from line 3 of Figure 9.3 that there are infinitely many solutions dependent on arbitrary constants b and d. Thus, we conclude that the matrices are linearly dependent. The second part of this problem has infinitely many correct solutions. If we choose the parameter d to be 1 and b to be 0, we obtain $-2\,\mathbf{U} + \mathbf{X} + \mathbf{Y} = 0$, so that $\mathbf{Y} = 2\mathbf{U} - \mathbf{X}$. (The parameters are substituted in the order listed in subs in line 3, and so, the value of b affects a, etc.) Any assignment of values for the parameters b and d will lead to a correct solution, except $b = d = 0$. (What's wrong with this choice?)

```
> U := matrix(4,2,[2,3,1,4,3,2,1,5]):
  V := matrix(4,2,[1,5,4,2,3,1,2,2]):
  X := matrix(4,2,[3,8,5,6,6,3,3,7]):
  Y := matrix(4,2,[1,-2,-3,2,0,1,-1,3]):
> eq := a*U + b*V + c*X + d*Y = 0;
```

$$eq := a\,U + b\,V + c\,X + d\,Y = 0$$

```
> matsolve(eq);
```

$$\{b = b, \quad d = d, \quad a = b - 2d, \quad c = -b + d\}$$

```
> subs(a=b-2*d, c=-b+d, b=0, d=1, eq);
```

$$-2\,U + X + Y = 0$$

Figure 9.3: | Testing elements in $\mathbb{M}_{4,2}$ for independence |

Solved Problem 3: In $C(-\infty, \infty)$ determine if the functions e^x, $\sin(x)$, $\cos(x)$, and $\ln(x^2 + 1)$ are linearly independent.

Solution: We wish to determine if there are any nontrivial solutions of

$$ae^x + b\sin(x) + c\cos(x) + d\ln(x^2 + 1) = 0$$

It is important to remember that the zero in this equation is the constant function $f(x) \equiv 0$ rather than the real number zero. Thus, we want to see if there are values of a, b, c, and d, not all zero, that make $ae^x + b\sin(x) + c\cos(x) + d\ln(x^2 + 1)$ zero for *all* x. If the expression is zero for *all* x, it must be zero for any four particular values of x: for example, $x = 0$, 1, 2, and 3. Why did we choose these four? Just because they are convenient. More is said about this at the end of the problem. We now seek solutions of the following system of equations:

$$\begin{aligned} a + c &= 0 \\ ae + b\sin(1) + c\cos(1) + d\ln(2) &= 0 \\ ae^2 + b\sin(2) + c\cos(2) + d\ln(5) &= 0 \\ ae^3 + b\sin(3) + c\cos(3) + d\ln(10) &= 0 \end{aligned}$$

It is tedious to enter these equations, so here's a way that saves typing. Define

```
f := x -> a*exp(x) + b*sin(x) + c*cos(x) + d*ln(x^2+1)
```

(See line 1 of Figure 9.4.) We can now enter the whole system of equations with

```
sys := {f(0), f(1), f(2), f(3)}
```

The solution to the system is displayed in line 3. We can conclude, because the only solution is the trivial one, that the four functions are linearly independent.

Maple Warning: To enter the function e^x, we must use `exp(x)` as we did in Figure 9.4. Maple interprets `e` as a variable just as it would `a` or `b`; the number e is denoted by `E` in Maple.

We chose $x = 0$, $x = 1$, $x = 2$, and $x = 3$ because they are convenient, but we could have picked any four distinct values and they would probably have led us to the same conclusion. However, it is possible that the resulting system may have a nontrivial solution. What would you conclude then? Be careful. (See Exploration and Discovery Problem 1(b).)

```
> f := x -> a*exp(x)+b*sin(x)+c*cos(x)+d*ln(x^2+1);
```

$$f := x \rightarrow a \exp(x) + b \sin(x) + c \cos(x) + d \ln\left(x^2 + 1\right)$$

```
> sys := {f(0), f(1), f(2), f(3)};
```

$$sys := \{a + c, \quad ae + b\sin(1) + c\cos(1) + d\ln(2), \quad ae^2 + b\sin(2)$$
$$+ c\cos(2) + d\ln(5), \quad ae^3 + b\sin(3) + c\cos(3) + d\ln(10)\}$$

```
> solve(sys);
```

$$\{a = 0, \quad b = 0, \quad c = 0, \quad d = 0\}$$

Figure 9.4: | Independence of functions in $C(-\infty, \infty)$ |

Solved Problem 4: Show that the set

$$S = \{(2, 6, 3, 4, 2), (3, 1, 5, 8, 3), (5, 1, 2, 6, 7), (8, 4, 3, 2, 6), (5, 5, 6, 3, 4)\}$$

is a basis for $\mathbb{R}^5$ and find the coordinate vector of $(3, 4, 1, 7, 8)$ relative to the ordered basis S.

Solution: There are several correct approaches to this problem but we will take the most efficient one. Define A to be the 5 by 6 matrix

$$A := \begin{bmatrix} 2 & 3 & 5 & 8 & 5 & 3 \\ 6 & 1 & 1 & 4 & 5 & 4 \\ 3 & 5 & 2 & 3 & 6 & 1 \\ 4 & 8 & 6 & 2 & 3 & 7 \\ 2 & 3 & 7 & 6 & 4 & 8 \end{bmatrix}$$

whose columns are the six given vectors (not shown in Figure 9.5). Now row-reduce this matrix as in line 1 of Figure 9.5. This solves both problems at once. Since the first five columns reduce to the 5 by 5 identity matrix, S is a basis for $\mathbb{R}^5$ and we see the desired coordinate vector relative to the ordered basis S as the last column in line 2 in Figure 9.5:

$$\left(\frac{557}{636}, -\frac{83}{106}, \frac{191}{106}, -\frac{499}{636}, \frac{55}{318}\right)$$

```
> rref(A);
```

$$\begin{bmatrix} 1 & 0 & 0 & 0 & 0 & \frac{557}{636} \\ 0 & 1 & 0 & 0 & 0 & \frac{-83}{106} \\ 0 & 0 & 1 & 0 & 0 & \frac{191}{106} \\ 0 & 0 & 0 & 1 & 0 & \frac{-499}{636} \\ 0 & 0 & 0 & 0 & 1 & \frac{55}{318} \end{bmatrix}$$

Figure 9.5: $\boxed{\text{A basis for } \mathbb{R}^5}$

Solved Problem 5: Show that $S = \{2-x+x^3,\ x+3x^2,\ 5-x^2-x^3,\ 2+x+2x^2+4x^3\}$ is a basis for the space $\mathbb{P}_3$ of polynomials of degree no more than 3. Find the coordinate vector of $1 + x + x^2 + x^3$ relative to the ordered basis S.

Solution: We will make use of the theorem that a set of n linearly independent vectors in a vector space of dimension n is a basis. Since $\mathbb{P}_3$ has dimension 4, we need only show that the four polynomials are linearly independent. Define each of the following as a separate function:

$$P := x \to 2 - x + x^3$$
$$Q := x \to x + 3x^2$$
$$R := x \to 5 - x^2 - x^3$$
$$S := x \to 2 + x + 2x^2 + 4x^3$$

Next, enter the expression $aP(x) + bQ(x) + cR(x) + dS(x)$ as **expr** in line 2 of Figure 9.6. Collect like powers of x. Since each coefficient must be zero, we see that $aP(x) + bQ(x) + cR(x) + dS(x) = 0$ if and only if a, b, c, and d are solutions of the following system of equations:

$$\begin{aligned} a - c + 4d &= 0 \\ 3b - c + 2d &= 0 \\ -a + b + d &= 0 \\ 2a + 5c + 2d &= 0 \end{aligned}$$

This system appears as line 4 of Figure 9.6. If we solve, we get line 5, which shows that $a = b = c = d = 0$ is the unique solution. Thus, the four functions are linearly independent. We could also use the other technique described in Solved Problem 3. If $aP(x) + bQ(x) + cR(x) + dS(x)$ is the zero function, then it must be 0 for *all* values of x. Use **subs** to substitute the values 0, 1, 2, and 3 for x in **expr**; then solve the resulting system. Although this method is often more laborious to execute by pencil and paper than the first, the reverse is true with Maple. To find the coordinates of $1 + x + x^2 + x^3$ relative to the ordered basis S, set its coefficients equal to those of $aP(x)+bQ(x)+cR(x)+dS(x)$. To simplify the calculations, we subtract $1+x+x^2+x^3$ from **expr** as seen in line 6, and then solve coefficient system in line 8.

We read the coordinate vector as $\left(-\frac{36}{79}, \frac{10}{79}, \frac{17}{79}, \frac{33}{79}\right)$. We have checked our work in line 9 of Figure 9.6.

```
>  P := x -> 2-x+x^3:
   Q := x -> x+3*x^2:
   R := x -> 5-x^2-x^3:
   S := x -> 2+x+2*x^2+4*x^3:
>  expr := a*P(x)+b*Q(x)+c*R(x)+d*S(x);
```

$$expr := a\left(2 - x + x^3\right) + b\left(x + 3x^2\right) + c\left(5 - x^2 - x^3\right) + d\left(2 + x + 2x^2 + 4x^3\right)$$

```
>  collect(expr,x);
```

$$(a - c + 4d)\, x^3 + (3b - c + 2d)\, x^2 + (-a + b + d)\, x + 2a + 5c + 2d$$

```
>  sys := {coeffs(",x)};
```

$$\{a - c + 4d, \quad 3b - c + 2d, \quad -a + b + d, \quad 2a + 5c + 2d\}$$

```
>  solve(sys);
```

$$\{a = 0, \quad b = 0, \quad c = 0, \quad d = 0\}$$

```
>  collect(expr-(1+x+x^2+x^3),x):
>  sys := {coeffs(",x)};
```

$$\{a - c + 4d - 1, \quad 3b - c + 2d - 1, \quad -a + b + d - 1, \quad 2a + 5c + 2d - 1\}$$

```
>  solve(sys);
```

$$\left\{a = \frac{-36}{79}, \quad b = \frac{10}{79}, \quad c = \frac{17}{79}, \quad d = \frac{33}{79}\right\}$$

```
>  subs(",expr);
```

$$1 + x + x^2 + x^3$$

Figure 9.6: A basis for $\mathbb{P}_3$

9.3   Exercises

1. Determine if the following sets are linearly independent. If they are linearly dependent write one element as a linear combination of the others. Unless otherwise instructed by your teacher, you may use any method presented in this chapter.

   (a) $\{(2,5,6,8,3,1),\ (5,3,9,1,4,2),\ (4,4,1,2,7,4),\ (8,9,4,1,3,5)\}$

   (b) $\{(4,2,3,4,6,7),\ (3,2,4,1,3,2),\ (2,1,1,3,1,2),\ (5,3,6,2,8,7)\}$

   (c) $\{\cos(x),\sin(x),\cos(2x),\sin(2x)\}$

   (d) $\{\cos^2(x),\cos(2x),8,e^x\}$

   (e) $\{(x-r)^2,(x-s)^2,(x-t)^2\}$, where r, s, and t are distinct numbers.

   (f) $\{\sin(x-1),\ \sin(x-2),\ \sin(x-3)\}$

   (g) $\{\sin(x),\ \sin(x-\pi),\ \sin(x-\frac{\pi}{2})\}$

   (h) $\{\ln(x^6+10x^4+33x^2+36),\ln(x^6+9x^4+27x^2+27),\ln(x^6+11x^4+40x^2+48)\}$

2. Find a basis for the subspace spanned by the given elements in Exercise 1.

3. In each of the following cases, show that the set S is a basis for the given vector space and find the coordinates of the given element $\mathbf{v}$ relative to S as an *ordered* basis.

   (a) $\mathbb{R}^5$;
   $S=\{(3,1,3,2,6),(4,5,7,2,4),(3,2,1,5,4),(2,9,1,4,4),(3,3,6,6,7)\}$;
   $\mathbf{v}=(2,4,1,2,3)$

   (b) $\mathbb{P}_3$;
   $S=\{x^3-x+1,\ x^3+x^2+3,\ 2x^3+3x^2-x+4,\ x^3+4x^2+5x-2\}$;
   $\mathbf{v}=3x^3-4x^2+2x-7$

   (c) $\mathbb{M}_{2,2}$;
   $$S=\left\{\begin{bmatrix}2&7\\3&9\end{bmatrix},\begin{bmatrix}3&5\\2&5\end{bmatrix},\begin{bmatrix}5&2\\9&7\end{bmatrix},\begin{bmatrix}3&6\\1&4\end{bmatrix}\right\};$$
   $$\mathbf{v}=\begin{bmatrix}4&2\\9&0\end{bmatrix}$$

4. In Solved Problem 5, show directly that the four polynomials span $\mathbb{P}_3$ by showing that the equation $f + g\,x + h\,x^2 + i\,x^3 = a\,p(x) + b\,q(x) + c\,r(x) + d\,s(x)$ has a solution for a, b, c, and d, no matter what the values of f, g, h, and i are.

5. Suppose $S = \{\mathbf{v}_1, \mathbf{v}_2, \mathbf{v}_3, \mathbf{v}_4\}$ is an ordered basis for $\mathbb{R}^4$ and that the coordinate vectors relative to S of $(2,3,5,1)$, $(3,7,2,4)$, $(5,1,1,3)$, and $(3,1,3,3)$, are respectively, $(1,5,2,3)$, $(4,1,2,2)$, $(2,2,7,6)$, and $(4,4,1,2)$. Find $\mathbf{v}_1$, $\mathbf{v}_2$, $\mathbf{v}_3$, and $\mathbf{v}_4$.

9.4   Exploration and Discovery

1. In Solved Problem 3 we concluded that the four functions e^x, $\sin(x)$, $\cos(x)$, and $\ln(x)$ are linearly independent by substituting the values 0, 1, 2, and 3 for x in the equation $a\,e^x + b\sin(x) + c\cos(x) + d\ln(x) = 0$ and solving the resulting system. In Solved Problem 5 we remarked that you could do the same for the four polynomials presented there.

   (a) Carry out the method of substituting the values 0, 1, 2, and 3 for x in the equation $a\,P(x) + b\,Q(x) + c\,R(x) + d\,S(x) = 0$ in Solved Problem 5 and solving the resulting system of equations.

   (b) At the end of Solved Problem 3, we asked what you would conclude if the system had nontrivial solutions. Let's try out our method on the three functions: $x+1$, $x^3 - 3x^2 + 3x + 1$, and $2 - \cos(\frac{\pi x}{2})$. Substitute the values 0, 1, and 2 for x in the equation $a(x+1) + b(x^3 - 3x^2 + 3x + 1) + c\left(2 - \cos(\frac{\pi x}{2})\right) = 0$ and solve the resulting system. Explain what happens.

   (c) Instead of substituting the values 0, 1, and 2 for x in the equation $a(x+1) + b(x^3 - 3x^2 + 3x + 1) + c\left(2 - \cos(\frac{\pi x}{2})\right) = 0$, try substituting 1, 2, and 3 for x. Explain what happens.

   (d) Try substituting 3, 4, and 5 for x. Explain what happens. Are $x + 1$, $x^3 - 3x^2 + 3x + 1$, and $2 - \cos(\frac{\pi x}{2})$ linearly independent? Summarize what you have concluded about this method.

2. **For students who have studied calculus.** There is a method for establishing independence of functions on an interval that arises in differential equations. It uses the *Wronskian*. We denote the ith derivative of f by $f^{(i)}$. If f_1, f_2, ..., f_n are functions on (a, b) that have derivatives of order $n - 1$, their Wronskian is defined as

$$W(f_1, f_2, \cdots, f_n) = \begin{vmatrix} f_1 & f_2 & \cdots & f_n \\ f_1' & f_2' & \cdots & f_n' \\ \vdots & \vdots & \ddots & \vdots \\ f_1^{(n-1)} & f_2^{(n-1)} & \cdots & f_n^{(n-1)} \end{vmatrix}$$

Theorem: If $W(f_1, f_2, \cdots, f_n)$ is not identically zero, then $f_1, f_2, \ldots, f_n$ are linearly independent.

Wronskian is defined in the `linalg` package as a matrix, so we need to define

$$W := \det @ \text{Wronskian}$$

The syntax is `W(A,x)`, where `A` is a list of functions and `x` is the independent variable.

(a) The theorem stated above is not difficult to prove. Try it. (<u>Hint</u>: Suppose that $f_1, f_2, \ldots, f_n$ are linearly dependent. Then show that their Wronskian must be zero.)

(b) Enter and simplify `W([exp(x), sin(x), cos(x), ln(x^2+1)], x)`. Compare your conclusion with the one obtained in Solved Problem 3.

(c) Use the Wronskian to test the independence of the functions in Solved Problem 5.

(d) Determine the value of the Wronskian for $\{1, x, x^2, \cdots, x^n\}$. (<u>Hint</u>: Consider several values of n. Then look at the ratio of the nth calculation to the $(n-1)$st.)

(e) Determine the value of the Wronskian for $\{e^x, e^{2x}, \cdots, e^{nx}\}$. (<u>Hint</u>: Consider several values of n. Then look at the ratio of the nth calculation to the $(n-1)$st.)

(f) Let $f_1 := x \rightarrow x^2$ and $f_2 := x \rightarrow x\,|x|$. Show that $W(f_1, f_2) \equiv 0$ on $(-1,+1)$, but f_1 and f_2 are linearly independent on $(-1,+1)$. How does this relate to the converse of the theorem?

9.5 Sets of Vectors

LABORATORY EXERCISE 9–1

Name _____   Due Date _____

Let $A = \begin{bmatrix} 3 & 5 & 9 & 8 \\ 2 & 3 & 1 & 7 \\ 4 & 1 & 2 & 9 \\ 6 & 3 & 4 & 1 \end{bmatrix}$. Let S denote the subspace of $\mathbb{R}^4$ spanned by the vectors $(3,2,1,1)$, $(1,1,3,1)$, $(2,1,-2,0)$, and $(4,3,4,2)$.

1. Find a spanning set for the nullspace of A.

2. Find a basis for S.

3. Find all vectors that are in both S and the nullspace of A.

4. The set in 3 above is a subspace of $\mathbb{R}^4$. Find a basis for it.

Chapter 10
ROW SPACE, COLUMN SPACE, AND NULLSPACE

LINEAR ALGEBRA CONCEPTS

- **Row space**
- **Column space**
- **Nullspace**
- **Rank**
- **Nullity**

10.1 Introduction

In this chapter we look at methods for producing bases for important vector spaces associated with matrices.

10.2 Solved Problems

Solved Problem 1: Find bases for (a) the row space, (b) the column space, and (c) the nullspace of the following matrix.

$$\begin{bmatrix} 1 & 5 & 2 & 4 & 4 & 7 \\ 3 & 2 & 4 & 9 & 1 & 3 \\ 5 & 2 & 4 & 8 & 5 & 7 \\ 9 & 9 & 10 & 21 & 10 & 17 \end{bmatrix}$$

<u>Solution to (a)</u>: Enter the 4 by 6 matrix, naming it A. We use the fact that *the nonzero rows of the reduced echelon form of A form a basis for the row space of A.* Thus, we simply row-reduce A with `rref`. From line 2 of Figure 10.1, we read a basis for the row space of A as

$$\left\{ \left(1,0,0,-\frac{1}{2},2,2\right), \left(0,1,0,-\frac{3}{16},\frac{9}{8},\frac{13}{8}\right), \left(0,0,1,\frac{87}{32},-\frac{29}{16},-\frac{25}{16}\right) \right\}$$

<u>Solution to (b)</u>: The column space of A is the row space of the transpose of A. Therefore, we may row-reduce A^t. From line 3 of Figure 10.1 we obtain the basis

$\{(1,0,0,1), (0,1,0,1), (0,0,1,1)\}$. Notice as well that the dimensions of the row and column spaces are the same. (This is always true; there is a theorem to that effect in your text.)

```
> A := matrix(4,6, [1,5,2,4,4,7,  3,2,4,9,1,3,  5,2,4,8,5,7,
    9,9,10,21,10,17]);
```

$$A := \begin{bmatrix} 1 & 5 & 2 & 4 & 4 & 7 \\ 3 & 2 & 4 & 9 & 1 & 3 \\ 5 & 2 & 4 & 8 & 5 & 7 \\ 9 & 9 & 10 & 21 & 10 & 17 \end{bmatrix}$$

```
> rref(A);
```

$$\begin{bmatrix} 1 & 0 & 0 & \frac{-1}{2} & 2 & 2 \\ 0 & 1 & 0 & \frac{-3}{16} & \frac{9}{8} & \frac{13}{8} \\ 0 & 0 & 1 & \frac{87}{32} & \frac{-29}{16} & \frac{-25}{16} \\ 0 & 0 & 0 & 0 & 0 & 0 \end{bmatrix}$$

```
> rref(transpose(A));
```

$$\begin{bmatrix} 1 & 0 & 0 & 1 \\ 0 & 1 & 0 & 1 \\ 0 & 0 & 1 & 1 \\ 0 & 0 & 0 & 0 \\ 0 & 0 & 0 & 0 \\ 0 & 0 & 0 & 0 \end{bmatrix}$$

Figure 10.1: Row and column spaces

There is an alternative method for finding a basis for the column space that does not require the row reduction of the transpose of A: *The columns of A corresponding to columns of the reduced form of A that contain leading "1"s form a basis for the column space.* From line 2 of Figure 10.1, the first three columns of the reduced form of A contain leading "1"s. Thus, the first three columns of A

$$\begin{bmatrix} 1 \\ 3 \\ 5 \\ 9 \end{bmatrix}, \begin{bmatrix} 5 \\ 2 \\ 2 \\ 9 \end{bmatrix}, \begin{bmatrix} 2 \\ 4 \\ 4 \\ 10 \end{bmatrix}$$

also form a basis for the column space of A. We emphasize that both $\{(1,0,0,1),$ $(0,1,0,1), (0,0,1,1)\}$ and $\{(1,3,5,9), (5,2,2,9), (2,4,4,10)\}$ are correct answers. It is interesting to notice that these two methods produce entirely different bases, but the subspaces spanned by the two bases are the same.

Solution to (c): The nullspace of A is the subspace of solutions of $A\mathbf{x} = 0$; thus we must solve this system of equations and describe the solution set. There are two ways to use Maple to assist us. We will present both.

Method 1 for (c): Using line 2 of Figure 10.1, we recover the solution *with pencil and paper* from the row-reduced form of A.

$$\begin{aligned}
x_4 &= r \\
x_5 &= s \\
x_6 &= t \\
x_1 &= \tfrac{1}{2}r - 2s - 2t \\
x_2 &= \tfrac{3}{16}r - \tfrac{9}{8}s - \tfrac{13}{8}t \\
x_3 &= -\tfrac{87}{32}t + \tfrac{29}{16}s + \tfrac{25}{16}t
\end{aligned}$$

To parlay this into a basis for the nullspace, set each parameter equal to 1 and the others to 0 in turn, as illustrated in the display below. The three column vectors form the desired basis.

$$\begin{matrix} r=1 \\ s=0 \\ t=0 \end{matrix} \implies \begin{aligned} x_1 &= \tfrac{1}{2} \\ x_2 &= \tfrac{3}{16} \\ x_3 &= -\tfrac{87}{32} \\ x_4 &= 1 \\ x_5 &= 0 \\ x_6 &= 0 \end{aligned} \implies \begin{bmatrix} \tfrac{1}{2} \\ \tfrac{3}{16} \\ -\tfrac{87}{32} \\ 1 \\ 0 \\ 0 \end{bmatrix}$$

$$\begin{matrix} r=0 \\ s=1 \\ t=0 \end{matrix} \implies \begin{aligned} x_1 &= -2 \\ x_2 &= -\tfrac{9}{8} \\ x_3 &= \tfrac{29}{16} \\ x_4 &= 0 \\ x_5 &= 1 \\ x_6 &= 0 \end{aligned} \implies \begin{bmatrix} -2 \\ -\tfrac{9}{8} \\ \tfrac{29}{16} \\ 0 \\ 1 \\ 0 \end{bmatrix}$$

$$
\begin{array}{c}
r = 0 \\
s = 0 \\
t = 1
\end{array}
\implies
\begin{array}{rcl}
x_1 & = & -2 \\
x_2 & = & -\frac{13}{8} \\
x_3 & = & \frac{25}{16} \\
x_4 & = & 0 \\
x_5 & = & 0 \\
x_6 & = & 1
\end{array}
\implies
\begin{bmatrix}
-2 \\
-\frac{13}{8} \\
\frac{25}{16} \\
0 \\
0 \\
1
\end{bmatrix}
$$

Method 2 for (c): The idea here is to obtain the solution of $Ax = 0$ with the variables displayed by Maple. Continuing from Figure 10.1, define the vector $T :=$ (x, y, z, u, v, w). Evaluate, as a matrix, the product AT (line 2). (Note that Maple automatically uses T as a column vector in this product.) Now convert the result to a set, then solve the system. Line 4 shows three parameters in the solution. To have the answer in the same form as before, solve again using the solve variables x, y, and z. Line 6, Figure 10.2, shows B, the result of substituting the solution into T.

```
> T := vector([x,y,z,u,v,w]):
> evalm(A &* T);
```

$$
[x + 5y + 2z + 4u + 7w \quad 3x + 2y + 4z + 9u + v + 3w
$$
$$
5x + 2y + 4z + 8u + 5v + 7w \quad 9x + 9y + 10z + 21u + 10v + 17w]
$$

```
> sys := convert('',set):
> solve(sys);
```

$$
\{u = \tfrac{16}{3}y + 6v + \tfrac{26}{3}w, \quad z = -\tfrac{29}{2}y - \tfrac{29}{2}v - 22w, \quad x = \tfrac{8}{3}y + v + \tfrac{7}{3}w,
$$
$$
v = v, \quad w = w, \quad y = y\}
$$

```
> soln := solve(sys, {x,y,z});
```

$$
soln := \left\{ y = \frac{3}{16}u - \frac{9}{8}v - \frac{13}{8}w, \quad z = -\frac{87}{32}u + \frac{29}{16}v + \frac{25}{16}w, \quad x = \frac{1}{2}u - 2v - 2w \right\}
$$

```
> B := subs(soln, T);
```

$$
B := \left[\tfrac{1}{2}u - 2v - 2w \quad \tfrac{3}{16}u - \tfrac{9}{8}v - \tfrac{13}{8}w \quad -\tfrac{87}{32}u + \tfrac{29}{16}v + \tfrac{25}{16}w \quad u \quad v \quad w \right]
$$

Figure 10.2: | Nullspace, second method: Part I |

Solved Problems

To generate the three nullspace basis vectors, Figure 10.3. Remember to use `evalm(B)` in the s~~itute for t~~ into the elements of B.   ~~ments s~~

> B1 := subs(u=1, v=0, w=0, evalm(B));

$$B1 := \begin{bmatrix} \frac{1}{2} & \frac{3}{16} & -\frac{87}{32} & 1 & 0 & 0 \end{bmatrix}$$

> B2 := subs(u=0, v=1, w=0, evalm(B));

$$B2 := \begin{bmatrix} -2 & -\frac{9}{8} & \frac{29}{16} & 0 & 1 & 0 \end{bmatrix}$$

> B3 := subs(u=0, v=0, w=1, evalm(B));

$$B3 := \begin{bmatrix} -2 & -\frac{13}{8} & \frac{25}{16} & 0 & 0 & 1 \end{bmatrix}$$

Figure 10.3: | Nullspace, second method: Part II |

Solved Problem 2: Verify that *rank + nullity = number of columns* if

$$A = \begin{bmatrix} 3 & 5 & 1 & 7 \\ 2 & 8 & 6 & 3 \\ 1 & 5 & 2 & 1 \\ 4 & 8 & 5 & 9 \end{bmatrix}$$

Solution: Enter the 4 by 4 matrix, naming it C, and then row-reduce. From line 2 of Figure 10.4 we see that $(1,0,0,\frac{91}{30})$, $(0,1,0,-\frac{13}{30})$, and $(0,0,1,\frac{1}{15})$ form a basis for the row space. Thus, the rank of A is 3.

Following the procedure outlined in Solved Problem 1, we obtain the single vector $\left(-\frac{91}{30}, \frac{13}{30}, \frac{1}{15}, 1\right)$ as a basis for the nullspace. Thus, the nullity is 1, and

$$rank + nullity = 3 + 1 = 4 = number\ of\ columns.$$

The reason why this formula works in general is clear from this example: The rank a matrix is the number of nonzero rows in the reduced form of A (which is the sa~~ as the number of columns containing leading "1"s), and the nullity is the numb~~

$k = 0$, which is the number of columns in the reduced

parameters in the soluti leading "1"s.
form of A that do not

```
> C := matrix(4    ,5,1,7, 2,8,6,3, 1,5,2,1, 4,8,5,9]);
```

$$C := \begin{bmatrix} 3 & 5 & 1 & 7 \\ 2 & 8 & 6 & 3 \\ 1 & 5 & 2 & 1 \\ 4 & 8 & 5 & 9 \end{bmatrix}$$

```
> rre  );
```

$$\begin{bmatrix} 1 & 0 & 0 & \frac{91}{30} \\ 0 & 1 & 0 & \frac{-13}{30} \\ 0 & 0 & 1 & \frac{1}{15} \\ 0 & 0 & 0 & 0 \end{bmatrix}$$

Figure 10.4: Rank and nullity of a matrix

parameters in the solution of $A\mathbf{x} = 0$, which is the number of columns in the reduced form of A that do *not* contain leading "1"s.

```
> C := matrix(4,4, [3,5,1,7, 2,8,6,3, 1,5,2,1, 4,8,5,9]);
```

$$C := \begin{bmatrix} 3 & 5 & 1 & 7 \\ 2 & 8 & 6 & 3 \\ 1 & 5 & 2 & 1 \\ 4 & 8 & 5 & 9 \end{bmatrix}$$

```
> rref(C);
```

$$\begin{bmatrix} 1 & 0 & 0 & \frac{91}{30} \\ 0 & 1 & 0 & \frac{-13}{30} \\ 0 & 0 & 1 & \frac{1}{15} \\ 0 & 0 & 0 & 0 \end{bmatrix}$$

Figure 10.4: Rank and nullity of a matrix

To generate the three nullspace basis vectors, we substitute for the parameters in Figure 10.3. Remember to use `evalm(B)` in the `subs` statements so as to substitute into the elements of B.

```
> B1 := subs(u=1, v=0, w=0, evalm(B));
```

$$B1 := \begin{bmatrix} \frac{1}{2} & \frac{3}{16} & -\frac{87}{32} & 1 & 0 & 0 \end{bmatrix}$$

```
> B2 := subs(u=0, v=1, w=0, evalm(B));
```

$$B2 := \begin{bmatrix} -2 & -\frac{9}{8} & \frac{29}{16} & 0 & 1 & 0 \end{bmatrix}$$

```
> B3 := subs(u=0, v=0, w=1, evalm(B));
```

$$B3 := \begin{bmatrix} -2 & -\frac{13}{8} & \frac{25}{16} & 0 & 0 & 1 \end{bmatrix}$$

Figure 10.3: | Nullspace, second method: Part II |

Solved Problem 2: Verify that *rank + nullity = number of columns* if

$$A = \begin{bmatrix} 3 & 5 & 1 & 7 \\ 2 & 8 & 6 & 3 \\ 1 & 5 & 2 & 1 \\ 4 & 8 & 5 & 9 \end{bmatrix}$$

<u>Solution</u>: Enter the 4 by 4 matrix, naming it C, and then row-reduce. From line 2 of Figure 10.4 we see that $(1,0,0,\frac{91}{30})$, $(0,1,0,-\frac{13}{30})$, and $(0,0,1,\frac{1}{15})$ form a basis for the row space. Thus, the rank of A is 3.

Following the procedure outlined in Solved Problem 1, we obtain the single vector $\left(-\frac{91}{30}, \frac{13}{30}, \frac{1}{15}, 1\right)$ as a basis for the nullspace. Thus, the nullity is 1, and

$$rank + nullity = 3 + 1 = 4 = number\ of\ columns.$$

reason why this formula works in general is clear from this example: The rank of ix is the number of nonzero rows in the reduced form of A (which is the same umber of columns containing leading "1"s), and the nullity is the number of

10.3   Exercises

1. For each of the following matrices, find bases for the row space, column space, and nullspace. In each case, verify that *rank* + *nullity* = *number of columns*.

   (a) $\begin{bmatrix} 2 & 3 & 8 & 4 & 7 \\ 5 & 1 & 3 & 9 & 6 \\ 2 & 4 & 8 & 5 & 2 \\ 5 & 0 & 3 & 8 & 11 \end{bmatrix}$

   (b) $\begin{bmatrix} 2 & 3 & 4 & 7 & 6 & 2 \\ 2 & 3 & 4 & 3 & 2 & 1 \\ 2 & 3 & 4 & 1 & 9 & 7 \\ 4 & 6 & 8 & 0 & 1 & 3 \\ 4 & 6 & 8 & 4 & 2 & 4 \end{bmatrix}$

   (c) $\begin{bmatrix} 5 & 1 & 1 & 8 & 2 & 3 & 5 \\ 10 & 2 & 2 & 4 & 1 & 5 & 4 \\ 5 & 1 & 1 & 9 & 1 & 1 & 5 \\ 10 & 2 & 2 & 4 & 5 & 6 & 3 \end{bmatrix}$

2. Find a basis for the subspace of $\mathbb{R}^5$ spanned by

$$(2,4,3,6,1), (3,1,5,7,2), (5,5,8,13,3), (-1,3,-2,-1,-1), (8,6,13,20,5)$$

   (Hint: This subspace is the row space of some matrix.)

3. Find a basis for the subspace of $\mathbb{R}^6$ each element of which is orthogonal to all three of the vectors

$$(1,2,4,6,7,2), (4,1,7,9,8,6), (4,1,5,2,3,4)$$

   (Hint: This subspace is the nullspace of some matrix.)

4. Find the intersection of the nullspaces of the matrices

$$\begin{bmatrix} 4 & 5 & 9 & 2 & 3 & 4 \\ 3 & 7 & 6 & 5 & 2 & 1 \end{bmatrix} \quad \text{and} \quad \begin{bmatrix} 2 & 6 & 3 & 5 & 9 & 2 \\ 1 & 2 & 1 & 2 & 3 & 4 \\ 3 & 1 & 2 & 6 & 9 & 1 \end{bmatrix}$$

   (Hint: Is the intersection the nullspace of a single matrix?)

10.4   Exploration and Discovery

Let S be the subspace of $\mathbb{R}^5$ spanned by $(3, 1, 5, 7, 2)$ and $(5, 5, 8, 13, 3)$.

1. Find a 4 by 5 matrix whose nullspace is S. Explain your method.

2. Find a 3 by 5 matrix whose nullspace is S. Explain your method.

3. Find a 2 by 5 matrix whose nullspace is S. Explain the difficulty encountered here.

4. Suppose you are given a subspace S of $\mathbb{R}^n$ of dimension k. If there is an m by n matrix whose nullspace is S, what can you say about m in general? Explain.

5. Investigate the Maple commands `colspace`, `nullspace`, and `rowspace`. How do their responses differ from your earlier work in Solved Problems 1 and 2?

10.5   Subspaces Associated with a Matrix

LABORATORY EXERCISE 10–1

Name _____   Due Date _____

Let $A = \begin{bmatrix} 7 & 0 & 1 & 8 & 2 & 3 & 5 \\ 5 & 1 & 2 & 4 & 1 & 0 & 4 \\ 3 & 3 & 1 & 9 & 1 & 2 & 1 \\ 9 & -2 & 2 & 3 & 2 & 1 & 8 \end{bmatrix}$.

1. Find a basis for the nullspace of A.

2. Multiply each basis vector above by A. Explain the meaning of your answers.

3. Find a basis for the row space of A.

4. Find a basis for the column space of A by row-reducing the transpose of A.

5. Find a basis for the column space of A by row-reducing A itself.

Chapter 11
INNER PRODUCT SPACES

```
┌─────────────────────────────────────────┐
│        LINEAR ALGEBRA CONCEPTS           │
│                                          │
│   • General inner products              │
└─────────────────────────────────────────┘
```

11.1 Introduction

If A is a given invertible n by n matrix, and $\mathbf{x}$ and $\mathbf{y}$ are vectors in $\mathbb{R}^n$, we will denote $(A\mathbf{x}) \cdot (A\mathbf{y})$ by $\langle \mathbf{x}, \mathbf{y} \rangle$. This is called the *inner product on $\mathbb{R}^n$ generated by A*; it has the familiar properties of the dot product. Indeed, if A is the identity matrix, then $\langle \mathbf{x}, \mathbf{y} \rangle = \mathbf{x} \cdot \mathbf{y}$. Some texts may define $\langle \mathbf{x}, \mathbf{y} \rangle$ by $(A\mathbf{x})^t(A\mathbf{y})$, which amounts to the same thing; however, since Maple will not accept this syntax*, we will use $(A\mathbf{x}) \cdot (A\mathbf{y})$, which is entered in Maple as `innerprod(A &* x, A &* y)`. (We can shorten the typing by defining a Maple *operator* as follows. Enter

```
'&.' := innerprod
```

Then instead of `innerprod(A &* x, A &* y)`, you may use `(A &* x) &. (A &* y)`.)

11.2 Solved Problems

Solved Problem 1: Let $\langle \mathbf{x}, \mathbf{y} \rangle$ denote the inner product on $\mathbb{R}^4$ generated by the matrix

$$A = \begin{bmatrix} 2 & 3 & 8 & 1 \\ 3 & 1 & 5 & 2 \\ 3 & 3 & 9 & 7 \\ 2 & 1 & 4 & 1 \end{bmatrix}$$

Let $\mathbf{u} = (2, 4, 3, 1)$ and $\mathbf{v} = (3, 6, 1, 4)$.

1. Calculate $\langle \mathbf{u}, \mathbf{v} \rangle$.

---

*Using several `evalm`'s can work; e.g., `evalm(transpose(evalm(A &* u)) &* evalm(A &* v))`

2. Calculate the length, $||\mathbf{u}||$, of $\mathbf{u}$ with respect to this inner product.

3. Find a nonzero vector that is orthogonal to $\mathbf{u}$ with respect to this inner product.

Solution: (1) Enter the matrix and name it A. Next define the vectors $\mathbf{u} = (2, 4, 3, 1)$ and $\mathbf{v} = (3, 6, 1, 4)$. Finally, enter `innerprod(A &* u, A &* v)`. (The alternative form is `(A &* u) &. ( A &* v)`.) The inner product appears in line 3 of Figure 11.1.

(2) Recall that $||\mathbf{u}|| = \sqrt{\langle \mathbf{u}, \mathbf{u} \rangle}$. Thus we enter `sqrt(innerprod(A &* u, A &* u))`. The length of $\mathbf{u}$ appears in line 4 of Figure 11.1.

```
> A := matrix(4,4,[2,3,8,1, 3,1,5,2, 3,3,9,7, 2,1,4,1]);
```

$$A := \begin{bmatrix} 2 & 3 & 8 & 1 \\ 3 & 1 & 5 & 2 \\ 3 & 3 & 9 & 7 \\ 2 & 1 & 4 & 1 \end{bmatrix}$$

```
> u := vector([2,4,3,1]);
  v := vector([3,6,1,4]);
```

$$u := \begin{bmatrix} 2 & 4 & 3 & 1 \end{bmatrix}$$
$$v := \begin{bmatrix} 3 & 6 & 1 & 4 \end{bmatrix}$$

```
> innerprod(A &* u, A &* v);
```

$$5980$$

```
> sqrt(innerprod(A &* u, A &* u));
```

$$\sqrt{5555}$$

```
> innerprod(A &* u, A &* vector([p,q,r,s]));
```

$$361\, p + 327\, q + 1015\, r + 480\, s$$

Figure 11.1: $\boxed{\text{An inner product on } \mathbb{R}^4 \text{ generated by a matrix}}$

(3) We want to find p, q, r, and s (not all zero) such that $\langle \mathbf{u}, (p, q, r, s) \rangle = 0$. Thus, we use `innerprod(A &* u, A &* vector([p,q,r,s])` as seen in line 5 of Figure 11.1. From the result, we see that there are infinitely many solutions, one of which is $p = 1$, $q = -\frac{361}{327}$, $r = 0$, $s = 0$, yielding the vector $(1, -\frac{361}{327}, 0, 0)$.

Solved Problem 2: | **For students who have studied calculus.** | If f and g are continuous on the closed interval $[a, b]$, we define the inner product

$$\langle f, g \rangle := \int_a^b f(x) g(x) \ dx$$

In this problem we will take $[a, b] = [-1, 1]$.

Find $\|\sin(x)\|$ and show that if $k \neq \pm 1$ is an integer, then $\sin(\pi x)$ and $\sin(k\pi x)$ are orthogonal.

Solution: $\|\sin(x)\| = \sqrt{< \sin(x), \sin(x) >} = \sqrt{\int_{-1}^{1} \sin^2(x) \ dx}$. Integrate `sin(x)^2` from -1 to 1 as done in line 1 of Figure 11.2. Use `evalf` of `sqrt` to approximate

$$\|\sin(x)\| = \sqrt{1 - \sin(1)\cos(1)} \approx 0.7384790360$$

For the second part of the exercise, calculate the integral $\int_{-1}^{1} \sin(\pi x) \ \sin(k\pi x) \ dx$ as done in line 3 of Figure 11.2. Since k is an integer and $k \neq \pm 1$, we see that this reduces to 0. Thus, the two functions are orthogonal.

```
> int(sin(x)^2, x=-1..1);
```

$$-\cos(1)\sin(1) + 1$$

```
> evalf(sqrt("));
```

$$.7384790360$$

```
> int(sin(Pi*x)*sin(k*Pi*x), x=-1..1);
```

$$-2\frac{\sin(k\pi)}{\pi(-1+k)(1+k)}$$

Figure 11.2: | An inner product on $C(-1, 1)$ |

Solved Problem 3: For students who have studied calculus. Using the inner product of Solved Problem 2, find a quadratic polynomial with leading coefficient 1 that is orthogonal to both $\sin(x)$ and $\cos(x)$.

Solution: If we are going to calculate an inner product several times, it is convenient to define it as a function. Define the function

$$\texttt{inp := (f,g) -> int(f*g, x=-1..1)}$$

as seen in Figure 11.3. Next we name a second-degree polynomial y using `y := x^2 + a*x + b`. We wish to find a and b so that both $\langle y, \sin(x) \rangle$ and $\langle y, \cos(x) \rangle$ are zero. Thus, we enter and solve the system $\{\texttt{inp(y,sin(x))}, \texttt{inp(y,cos(x))}\}$. The result displayed in line 4 of Figure 11.3 tells us that the required polynomial is

$$y = x^2 + 1 - 2\cot(1)$$

```
> inp := (f,g) -> int(f*g, x=-1..1);
```

$$inp := (f, g) \rightarrow \int_{-1}^{1} fg \; dx$$

```
> y := x^2 + a*x + b:
> sys := {inp(y,sin(x)),inp(y,cos(x))};
```

$$sys := \{2a\sin(1) - 2a\cos(1), \quad 2b\sin(1) + 4\cos(1) - 2\sin(1)\}$$

```
> solve(sys);
```

$$\left\{ a = 0, \quad b = -\frac{2\cos(1) - \sin(1)}{\sin(1)} \right\}$$

Figure 11.3: A quadratic orthogonal to $\sin(x)$ and $\cos(x)$

11.3   Exercises

1. This problem refers to the inner product from Solved Problem 1. Be careful to use it and not the usual dot product. You may find it convenient to enter the matrix A and then enter inp := (x,y) -> innerprod(A &* x, A &* y) to define a function for the inner product.

   (a) Find $\langle (2,5,3,4),(7,1,4,2) \rangle$.

   (b) Find the angle (in radians) between $(2,5,3,4)$ and $(7,1,4,2)$ for this inner product. (<u>Hint</u>: The cosine of the angle between **u** and **v** is $\dfrac{<\mathbf{u},\mathbf{v}>}{\|\mathbf{u}\| \, \|\mathbf{v}\|}$.)

   (c) Find a unit vector in the direction of $(3,6,1,3)$.

   (d) Find all vectors simultaneously orthogonal to $(2,1,5,3)$, $(3,1,2,2)$, and $(2,4,3,5)$.

2. $\boxed{\textbf{For students who have studied calculus.}}$ Let $\langle f,g \rangle := \int_{-1}^{1} f(x)g(x)\, dx$.

   (a) Show that if p and q are positive integers and $p \neq q$, then $\|\cos(p\pi x)\| = 1$ and that $\cos(p\pi x)$ is orthogonal to $\cos(q\pi x)$.

   (b) Find a polynomial in $\mathbb{P}_2$ of length 1 that is orthogonal to both e^x and e^{-x}.

3. Let $\mathbf{a} = (1,2,3,2)$, $\mathbf{b} = (1,3,2,4)$, and $\mathbf{c} = (3,2,2,1)$. Suppose that $\langle \mathbf{u},\mathbf{v} \rangle$ is an inner product on $\mathbb{R}^4$ and that the following are known.

   $$\langle \mathbf{a},\mathbf{a} \rangle = 2, \quad \langle \mathbf{a},\mathbf{b} \rangle = 3, \quad \langle \mathbf{a},\mathbf{c} \rangle = 1, \quad \langle \mathbf{b},\mathbf{b} \rangle = 1, \quad \langle \mathbf{b},\mathbf{c} \rangle = 4, \quad \langle \mathbf{c},\mathbf{c} \rangle = 2$$

   Find $\langle (-1,3,3,5),(3,1,3,-1) \rangle$. (<u>Hint</u>: Can the given vectors be written as linear combinations of **a**, **b**, and **c**?)

4. $\boxed{\textbf{For students who have studied calculus.}}$ Compute the inner product

   $$\int_{-1}^{1} (ax+b)(cx+d)\, dx$$

   Find a matrix A such that the inner product $\langle (a,b),(c,d) \rangle$ that it generates on $\mathbb{R}^2$ equals $\int_{-1}^{1}(ax+b)(cx+d)\, dx$. (<u>Hint</u>: Assume A is diagonal: $\begin{bmatrix} r & 0 \\ 0 & s \end{bmatrix}$. For more details, see Exploration and Discovery Problem 2.)

11.4 Exploration and Discovery

1. Let A be an n by n matrix and let $\langle \mathbf{u}, \mathbf{v} \rangle = (A\mathbf{x}) \cdot (A\mathbf{y})$ be the inner product on $\mathbb{R}^n$ generated by A. The first three are not computer problems.

  (a) Show that $S = A^t A$ is symmetric, that is, $S^t = S$.

  (b) Show that $\langle \mathbf{x}, \mathbf{y} \rangle = \mathbf{x}^t S \mathbf{y}$. (Recall that $(A\mathbf{x}) \cdot (A\mathbf{y}) = (A\mathbf{x})^t (A\mathbf{y})$.)

  (c) In (b) above we see that the inner product can be expressed as $\langle \mathbf{x}, \mathbf{y} \rangle = \mathbf{x}^t S \mathbf{y}$, where S is symmetric. Let $S = \begin{bmatrix} 1 & 2 \\ 2 & 1 \end{bmatrix}$ and $\mathbf{x} = \begin{bmatrix} 1 \\ -1 \end{bmatrix}$. Find $\mathbf{x}^t S \mathbf{x}$. Discuss the conclusions you draw from this.

  (d) Let $\mathbf{u} = (1,2)$ and $\mathbf{v} = (3,4)$. Find a matrix A such that $\langle \mathbf{u}, \mathbf{v} \rangle = 0$.

  (e) Let $\mathbf{u} = (1,2)$ and $\mathbf{v} = (2,-1)$. Find a matrix A, other than the identity, such that $\langle \mathbf{u}, \mathbf{v} \rangle = 0$.

  (f) Describe all 2 by 2 matrices A that "preserve orthogonality;" that is, if $\mathbf{u} \cdot \mathbf{v} = 0$, then $\langle \mathbf{u}, \mathbf{v} \rangle = 0$. (Hint: Consider the standard basis vectors.)

  (g) Describe all n by n matrices A that "preserve orthogonality."

2. $\boxed{\textbf{For students who have studied calculus.}}$ Do Exercise 4.

  (a) Compute the inner product $\int_{-1}^{1}(ax^2 + bx + c)(dx^2 + ex + f)\,dx$.

  (b) Show that there is no diagonal matrix $\begin{bmatrix} r & 0 & 0 \\ 0 & s & 0 \\ 0 & 0 & t \end{bmatrix}$ such that the inner product $\langle (a,b,c), (d,e,f) \rangle$ that it generates on $\mathbb{R}^3$ equals

$$\int_{-1}^{1}(ax^2 + bx + c)(dx^2 + ex + f)\,dx$$

(Hint: Assume there is such a matrix and arrive at a contradiction.)

  (c) Find a matrix A such that the inner product $\langle (a,b,c), (d,e,f) \rangle$ that it generates on $\mathbb{R}^3$ equals $\int_{-1}^{1}(ax^2 + bx + c)(dx^2 + ex + f)\,dx$. (Hint: Try assuming that A has the form $\begin{bmatrix} r & 0 & 0 \\ 0 & s & 0 \\ u & 0 & t \end{bmatrix}$.)

  (d) Explore the situation for $\mathbb{P}_3$ and $\mathbb{R}^4$.

11.5   Inner Products

LABORATORY EXERCISE 11–1

Name _____   Due Date _____

Let $\langle \mathbf{x}, \mathbf{y} \rangle_A$ denote the inner product on $\mathbb{R}^3$ generated by $A := \begin{bmatrix} 2 & 3 & 1 \\ 1 & 5 & 2 \\ 3 & 1 & 1 \end{bmatrix}$. Let $\mathbf{u} = (1, 3, 1)$ and $\mathbf{v} = (2, 1, 1)$.

1. Calculate $\langle \mathbf{u}, \mathbf{v} \rangle_A$.

2. Calculate the lengths of $\mathbf{u}$ and $\mathbf{v}$ with respect to this inner product.

3. Find a nonzero vector that is orthogonal to $\mathbf{u}$ with respect to this inner product.

4. Find a nonzero vector that is simultaneously orthogonal to $\mathbf{u}$ and $\mathbf{v}$ with respect to this inner product.

Chapter 12
ORTHONORMAL BASES AND THE GRAM-SCHMIDT PROCESS

> *LINEAR ALGEBRA CONCEPTS*
>
> - **Orthonormal basis**
> - **Gram-Schmidt process**

12.1 Introduction

An orthonormal basis for an inner product space behaves very much like the standard basis for $\mathbb{R}^n$. In particular, coordinate vectors are easy to compute in terms of the inner product. The Gram-Schmidt process is a method by which we may construct such a basis for a finite dimensional inner product space. It is often tedious to carry out with pencil and paper, but Maple makes it easy.

12.2 Solved Problems

Solved Problem 1: Verify that

$$S := \left\{ \left(\tfrac{1}{\sqrt{7}}, \tfrac{2}{\sqrt{7}}, \tfrac{1}{\sqrt{7}}, \tfrac{1}{\sqrt{7}} \right), \left(\tfrac{-1}{\sqrt{42}}, \tfrac{-2}{\sqrt{42}}, \tfrac{-1}{\sqrt{42}}, \tfrac{6}{\sqrt{42}} \right), \left(\tfrac{-1}{\sqrt{30}}, \tfrac{-2}{\sqrt{30}}, \tfrac{5}{\sqrt{30}}, 0 \right), \left(\tfrac{2}{\sqrt{5}}, \tfrac{-1}{\sqrt{5}}, 0, 0 \right) \right\}$$

is an orthonormal basis for $\mathbb{R}^4$ and find $[(1,2,3,4)]_S$ (the coordinate vector of $(1,2,3,4)$ relative to the ordered basis S).

Solution: Enter and name the four vectors in S as follows. (See Figure 12.1.)

```
u1:=vector([1,2,1,1])/sqrt(7)        u2:=vector([-1,-2,-1,6])/sqrt(42)
u3:=vector([-1,-2,5,0])/sqrt(30)     u4:=vector([2,-1,0,0])/sqrt(5)
```

The first thing that might occur to us is to check the lengths and all possible inner products of these four vectors one by one. Maple can help us do that, but it is very laborious and there is a much more efficient way. We will enter the matrix whose rows are the four vectors and multiply it by its own transpose. Enter the expression

`A := stack(u1,u2,u3,u4)`. The result is displayed in line 2 of Figure 12.1. We've used `stack` to place the vectors as rows rather than using `augment` which would place the vectors in A's columns. To calculate the product of A with its transpose, use `evalm(A &* transpose(A))`. We see in line 3 of Figure 12.1 that the product is the identity matrix. (It is left to the reader to explain why we may then conclude that the four vectors form an orthonormal basis.)

```
> u1 := vector([1,2,1,1])/sqrt(7):
  u2 := vector([-1,-2,-1,6])/sqrt(42):
  u3 := vector([-1,-2,5,0])/sqrt(30):
  u4 := vector([2,-1,0,0])/sqrt(5):
> A := stack(u1,u2,u3,u4);
```

$$A := \begin{bmatrix} \frac{1}{7}\sqrt{7} & \frac{2}{7}\sqrt{7} & \frac{1}{7}\sqrt{7} & \frac{1}{7}\sqrt{7} \\ -\frac{1}{42}\sqrt{42} & -\frac{1}{21}\sqrt{42} & -\frac{1}{42}\sqrt{42} & \frac{1}{7}\sqrt{42} \\ -\frac{1}{30}\sqrt{30} & -\frac{1}{15}\sqrt{30} & \frac{1}{6}\sqrt{30} & 0 \\ \frac{2}{5}\sqrt{5} & -\frac{1}{5}\sqrt{5} & 0 & 0 \end{bmatrix}$$

```
> evalm(A &* transpose(A));
```

$$\begin{bmatrix} 1 & 0 & 0 & 0 \\ 0 & 1 & 0 & 0 \\ 0 & 0 & 1 & 0 \\ 0 & 0 & 0 & 1 \end{bmatrix}$$

Figure 12.1: $\boxed{\text{An orthonormal basis for } \mathbb{R}^4}$

For the second part of the problem we use the fact that the coordinate vector of $\mathbf{b}$ with respect to the ordered orthonormal basis $\{\mathbf{u}_1, \mathbf{u}_2, \mathbf{u}_3, \mathbf{u}_4\}$ is given by the following.

$$[\mathbf{b}]_S = (\mathbf{b} \cdot \mathbf{u}_1, \mathbf{b} \cdot \mathbf{u}_2, \mathbf{b} \cdot \mathbf{u}_3, \mathbf{b} \cdot \mathbf{u}_4)$$

Define $\mathbf{b} := (1,2,3,4)$. Since the ith coordinate is given by $\mathbf{b} \cdot \mathbf{u}_i$, it can be calculated with `dotprod(b,ui)`. All the coordinates of $[\mathbf{b}]_S$ can be calculated by using Maple's generator notation for vectors, `vector(length, i → ith component)`. The coordinate vector appears in line 3 of Figure 12.2. Notice that in line 4 we have calculated the product, $A\mathbf{b}$, and that this also gives the proper coordinate vector. Why does this work?

```
> b := vector([1,2,3,4]):
> b1 := dotprod(b,u1);
```

$$\frac{12}{7}\sqrt{7}$$

```
> vector(4, i -> dotprod(b,u.i));
```

$$\left[\frac{12}{7}\sqrt{7} \quad \frac{8}{21}\sqrt{42} \quad \frac{1}{3}\sqrt{30} \quad 0\right]$$

```
> evalm(A &* b);
```

$$\left[\frac{12}{7}\sqrt{7} \quad \frac{8}{21}\sqrt{42} \quad \frac{1}{3}\sqrt{30} \quad 0\right]$$

Figure 12.2: $\boxed{\text{Coordinates relative to a basis}}$

Solved Problem 2: Define the inner product for f and g in $C(-1,1)$ by

$$\langle f, g \rangle = \int_{-1}^{1} f(x)g(x)\,dx$$

Verify that $S = \left\{\frac{1}{\sqrt{2}}, \sin(\pi x), \sin(2\pi x), \sin(3\pi x)\right\}$ is an orthonormal set with respect to this inner product. Find the projection of $f(x) = x$ onto the subspace W spanned by these four vectors.

Solution: To assist in computing inner products, define the function

$$\text{ip} := (f,g) \rightarrow \text{int}(f*g, x=-1..1)$$

as we did in Chapter 11. (See Figure 12.3.) Enter and name the four vectors as $u1 := 1/\sqrt{2}$, $u2 := \sin(\pi x)$, $u3 := \sin(2\pi x)$, and $u4 := \sin(3\pi x)$. To test for orthonormality we must calculate the inner product of all pairs of these vectors. We have calculated `ip(u1,u1)` and `ip(u1,u2)` in lines 3 and 4 of Figure 12.3. We leave it to the reader to complete the process.

Since S is orthonormal, the projection of $\mathbf{v}$ onto W is given by

$$\langle \mathbf{v}, \mathbf{u}_1 \rangle \, \mathbf{u}_1 + \langle \mathbf{v}, \mathbf{u}_2 \rangle \, \mathbf{u}_2 + \langle \mathbf{v}, \mathbf{u}_3 \rangle \, \mathbf{u}_3 + \langle \mathbf{v}, \mathbf{u}_4 \rangle \, \mathbf{u}_4$$

Thus, we enter

$$F := ip(x,u1)*u1 + ip(x,u2)*u2 + ip(x,u3)*u3 + ip(x,u4)*u4.$$

The result is displayed in line 5 of Figure 12.3.

Recall that the projection of a vector $\mathbf{v}$ onto a subspace W may be thought of as the best approximation of $\mathbf{v}$ by a vector in W, in the sense that among all elements $\mathbf{w}$ in W, $\|\mathbf{v} - \mathbf{w}\|$ is minimal when $\mathbf{w}$ is the projection of $\mathbf{v}$ onto W.

```
> ip := (f,g) -> int(f*g, x=-1..1);
```

$$ip := (f,g) \rightarrow \int_{-1}^{1} f\,g\,dx$$

```
> u1 := 1/sqrt(2):   u2 := sin(Pi*x):   u3 := sin(2*Pi*x):   u4 := sin(3*Pi*x):
> ip(u1,u1);
```
$$1$$

```
> ip(u1,u2);
```
$$0$$

```
> F := ip(x,u1)*u1 + ip(x,u2)*u2 + ip(x,u3)*u3 + ip(x,u4)*u4;
```

$$F := 2\frac{\sin\left(\pi x\right)}{\pi} - \frac{\sin\left(2\pi x\right)}{\pi} + \frac{2}{3}\frac{\sin\left(3\pi x\right)}{\pi}$$

Figure 12.3: $\boxed{\text{The projection of } x \text{ onto a subspace of } C(-1,1)}$

The graph of x along with its projection onto W appears in Figure 12.4. The graph is made by plotting the set of expressions {x,F}. It should be noted how the graphs appear close together on $(-1,1)$ until we approach the endpoints. We do not expect the approximation to be good outside the interval $(-1,1)$, and this is shown clearly in the plot.

```
> plot({x,F}, x=-5/2..5/2,-3/2..3/2);
```

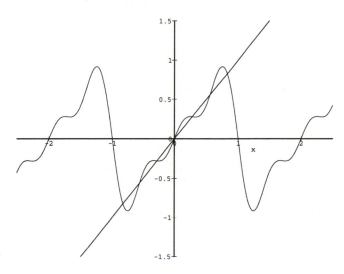

Figure 12.4: $\boxed{\text{Approximating } x \text{ on } (-1, 1) \text{ by sine functions}}$

Solved Problem 3: Find an orthonormal basis for the subspace of $\mathbb{R}^5$ spanned by the vectors $\mathbf{u}_1 = (1, 3, 2, 5, 2), \mathbf{u}_2 = (2, 1, 3, 2, 2), \mathbf{u}_3 = (1, 2, 3, 2, 1)$.

<u>Solution</u>: Define the vectors u1, u2, and u3 (not shown in Figure 12.5). The Gram-Schmidt process is tedious to carry out with pencil and paper, but Maple makes it easy. To avoid a lot of typing, it is helpful to create a function for the projection of $\mathbf{u}$ onto $\mathbf{v}$: define $\text{proj} := (\mathbf{u}, \mathbf{v}) \to \dfrac{\mathbf{u} \cdot \mathbf{v}}{\mathbf{v} \cdot \mathbf{v}} \mathbf{v}$ by

$$\text{proj} := (\text{u,v}) \; \text{->} \; \text{innerprod(u,v)/innerprod(v,v)*v}$$

for later use as done in Figure 12.5.

Now enter, in order,

$$v_1 := u_1$$
$$v_2 := u_2 - \text{proj}(u_2, v_1)$$
$$v_3 := u_3 - \text{proj}(u_3, v_1) - \text{proj}(u_3, v_2)$$

as separate expressions as shown in Figure 12.5. It is essential that vectors $\mathbf{v}_1$, $\mathbf{v}_2$, and $\mathbf{v}_3$ be defined in the order shown in Figure 12.5 (Why is this essential?). The vectors $\mathbf{v}_1$, $\mathbf{v}_2$, and $\mathbf{v}_3$ thus far defined form an *orthogonal* basis. We finish by normalizing each vector: for each $\mathbf{v}_i$, enter $\mathbf{v}_i := \mathbf{v}_i / |\mathbf{v}_i|$ as in `v1 := evalm(v1/norm(v1,2))`. (Recall that `norm(u,2)` gives the Euclidean length of a vector.)

The required orthonormal basis appears in line 5 of Figure 12.5. We will not do so here, but you should use Maple to check this answer by showing that each $\mathbf{v}_i$ is a unit vector and that they are mutually orthogonal. (Recall how we did this efficiently with a matrix in Solved Problem 1.)

```
> proj := (u,v) -> innerprod(u,v)/innerprod(v,v) * v;
```

$$proj := (u,v) \to \frac{innerprod\,(u,v)\,v}{innerprod(v,v)}$$

```
> v1 := u1;
```

$$v1 := u1$$

```
> v2 := u2 - proj(u2,v1);
```

$$v2 := u2 - \frac{25}{43}\,u1$$

```
> v3 := u3 - proj(u3,v1) - proj(u3,v2);
```

$$v3 := u3 - \frac{25}{107}\,u1 - \frac{64}{107}\,u2$$

```
>  v1:=evalm(v1/norm(v1,2));
   v2:=evalm(v2/norm(v2,2));
   v3:=evalm(v3/norm(v3,2));
```

$$v1 := \left[\tfrac{1}{43}\sqrt{43} \quad \tfrac{3}{43}\sqrt{43} \quad \tfrac{2}{43}\sqrt{43} \quad \tfrac{5}{43}\sqrt{43} \quad \tfrac{2}{43}\sqrt{43}\right]$$
$$v2 := \left[\tfrac{61}{13803}\sqrt{13803} \quad -\tfrac{32}{13803}\sqrt{13803} \quad \tfrac{79}{13803}\sqrt{13803} \quad -\tfrac{13}{4601}\sqrt{13803} \quad \tfrac{12}{4601}\sqrt{13803}\right]$$
$$v3 := \left[-\tfrac{23}{1284}\sqrt{321} \quad \tfrac{25}{856}\sqrt{321} \quad \tfrac{79}{2568}\sqrt{321} \quad -\tfrac{13}{856}\sqrt{321} \quad -\tfrac{71}{2568}\sqrt{321}\right]$$

Figure 12.5: Executing the Gram-Schmidt process in $\mathbb{R}^5$

Solved Problem 4: Using the inner product on $C(-1,1)$ defined by

$$\langle f, g \rangle := \int_{-1}^{1} f(x)g(x) \, dx$$

find an orthonormal basis for the subspace spanned by $\{x, x^2, x^3\}$.

Solution: This may appear to be more difficult than Solved Problem 3, but with Maple to do the calculations, the execution of the Gram-Schmidt process is the same in *any* inner product space. As before, to avoid a lot of typing, it is helpful to define the functions for inner product, projection, and vector length: enter

```
ip := (f,g) -> int(f*g, x=-1..1)
proj:= (u,v) -> ip(u,v)/ip(v,v)*v
```

and

```
vlength := u -> sqrt(ip(u,u))
```

for later use. (See Figure 12.6.) The reason for defining `vlength` is to avoid problems generated by `norm`'s introduction of absolute values.

After we define the three vectors as $u_1 := x$, $u_2 := x^2$, and $u_3 := x^3$ (not shown in Figure 12.6), we execute the Gram-Schmidt process exactly as in Solved Problem 1 using the projection function. Enter

```
v1 := u1
v2 := u2 - proj(u2,v1)
```

and

```
v3 := u3 - proj(u3,v1) - proj(u3,v2).
```

Now we can change the orthogonal basis $\{v_1, v_2, v_3\}$ to an orthonormal basis using $v_i := v_i/|v_i|$. Do this to obtain the orthonormal basis displayed in line 3 of Figure 12.6.

```
> ip := (f,g) -> int(f*g, x=-1..1):
  proj:= (u,v) -> ip(u,v)/ip(v,v)*v:
  vlength := u -> sqrt(ip(u,u)):

> v1 := u1;
  v2 := u2-proj(u2,v1);
  v3 := u3-proj(u3,v1)-proj(u3,v2);
```

$$v1 := x$$
$$v2 := x^2$$
$$v3 := x^3 - \tfrac{3}{5}x$$

```
> v1 := v1/vlength(v1);
  v2 := v1/vlength(v1);
  v3 := v1/vlength(v1);
```

$$v1 := \tfrac{1}{2}\,x\sqrt{6}$$
$$v2 := \tfrac{1}{2}\,x^2\sqrt{10}$$
$$v3 := \tfrac{5}{4}\left(x^3 - \tfrac{3}{5}x\right)\sqrt{14}$$

Figure 12.6: $\boxed{\text{The Gram-Schmidt process in } C(-1,1)}$

12.3   Exercises

1. Show that

$$S := \left\{ \left(\tfrac{1}{\sqrt{3}}, \tfrac{1}{\sqrt{3}}, 0, \tfrac{1}{\sqrt{3}} \right), \left(\tfrac{-1}{\sqrt{15}}, \tfrac{-1}{\sqrt{15}}, \tfrac{1}{\sqrt{15}}, \tfrac{2}{\sqrt{15}} \right), \left(\tfrac{2}{\sqrt{15}}, \tfrac{-1}{\sqrt{15}}, \tfrac{-1}{\sqrt{15}}, \tfrac{1}{\sqrt{15}} \right), \left(\tfrac{1}{\sqrt{3}}, 0, \tfrac{1}{\sqrt{3}}, \tfrac{-1}{\sqrt{3}} \right) \right\}$$

   is an orthonormal basis for $\mathbb{R}^4$ and find the coordinates of $(1,2,3,4)$ relative to the ordered basis S.

2. Find an orthonormal basis for the subspace of $\mathbb{R}^5$ spanned by

$$\{(1,0,2,3,1), \ (1,0,2,1,3), \ (1,2,3,1,0)\}.$$

3. Find an orthonormal basis for the nullspace of $\begin{bmatrix} 2 & 3 & 1 & 4 & 1 & 1 \\ 2 & 4 & 1 & 2 & 1 & 2 \\ 2 & 3 & 1 & 1 & 2 & 2 \end{bmatrix}$.

4. Let $\langle \mathbf{u}, \mathbf{v} \rangle$ be the inner product on $\mathbb{R}^4$ generated by $\begin{bmatrix} 1 & 2 & 1 & 3 \\ 2 & 1 & 3 & 2 \\ 1 & 1 & 1 & 2 \\ 3 & 2 & 1 & 3 \end{bmatrix}$. (See Chapter 11.) Find an orthonormal basis for $\mathbb{R}^4$ using this inner product by applying the Gram-Schmidt process to the standard basis for $\mathbb{R}^4$.

5. This is a variation of Solved Problem 2, where $\langle f, g \rangle = \int_{-1}^{1} f(x)g(x) \, dx$.

   (a) Show that $\left\{ \tfrac{1}{\sqrt{2}}, \cos(\pi x), \cos(2\pi x), \cos(3\pi x) \right\}$ is an orthonormal set.

   (b) Find the projection of x^2 onto the subspace spanned by these four functions.

   (c) On the same axes, graph x^2 and the projection of x^2 onto the subspace spanned by the new set S.

   (d) Find the projection of x onto the subspace spanned by these four functions. By examining the integrals, explain how you can predict the projections of x, x^3, x^5, ... without computing anything.

6. Let $\langle f, g \rangle = \int_{-1}^{1} f(x)g(x) \, dx$. Using this inner product, find an orthonormal basis for the subspace of $C(-1,1)$ spanned by $\sin^2(\pi x)$, $\sin^2(2\pi x)$, and $\sin^2(3\pi x)$.

12.4   Exploration and Discovery

1. Try to produce a Maple function that will execute the Gram-Schmidt process in a single step. (You can look into many Maple functions by setting `interface(verboseproc=2)` and then entering `print(`*name*`)` as in `print(` `linalg[GramSchmidt])`. Functions from packages need their "long form" name in the `print` statement.)

2. This is a continuation of Solved Problem 2 where we found the projection of $f(x) = x$ onto the subspace spanned by S.

   (a) Find the distance between $f(x) = x$ and its projection.

   (b) Include $\sin(4\pi x)$ in the set S and then find the projection of $f(x) = x$ onto the subspace spanned by this new set.

   (c) Find the distance between $f(x) = x$ and its new projection. Is the approximation better than before? Explain.

   (d) On the same axes, graph x and the projection of x onto the subspace spanned by the new set S.

   (e) Repeat all the steps above for $\sin(5\pi x)$.

   (f) What do you think will happen if you continue to include $\sin(6\pi x)$, $\sin(7\pi x)$, and so on in the set S?

12.5   Executing the Gram-Schmidt Process

LABORATORY EXERCISE 12–1

Name _____   Due Date _____

1. Find an orthonormal basis for the subspace of $\mathbb{R}^5$ spanned by

$$\{(0,3,3,1,1),\ (1,0,0,1,2),\ (1,2,5,7,8)\}$$

Carry out the procedure as we did in Solved Problem 3. Show the steps involved clearly.

2. Express each vector in the spanning set above as a linear combination of the orthonormal basis vectors you found.

Chapter 13
CHANGE OF BASIS AND ORTHOGONAL MATRICES

LINEAR ALGEBRA CONCEPTS

- **Transition (or change of basis) matrix**
- **Orthogonal matrice**

13.1   Introduction

If B is an ordered basis for a vector space V, then a vector $\mathbf{w}$ in V has a coordinate vector relative to B, which we will denote by $[\mathbf{w}]_B$. If B and B' are two ordered bases for V, then $\mathbf{w}$ has a coordinate vector relative to both, and one can be obtained from the other by multiplication by a matrix called the *transition matrix* or *change of basis matrix.*

If V is an inner product space and if B and B' are orthonormal ordered bases, then the transition matrix has a special structure and is called an *orthogonal* matrix.

Theorem: For an n by n matrix A, the following are equivalent.

1. A is orthogonal.

2. $A A^t = I$.

3. The rows of A form an orthonormal basis for $\mathbb{R}^n$.

4. The columns of A form an orthonormal basis for $\mathbb{R}^n$.

5. $||A\mathbf{v}|| = ||\mathbf{v}||$ for all $\mathbf{v}$ in $\mathbb{R}^n$.

We used the equivalence of 2 and 3 above in Solved Problem 1 of Chapter 12. Number 5 above says that A preserves distance.

Since Maple's `solve` does not handle matrix equations, we will define the function `matsolve` as we have done before (see Appendix B). After loading the `linalg` package, define the function

```
matsolve := eq -> solve(convert(evalm(lhs(eq)-rhs(eq)), set))
```

13.2   Solved Problems

Solved Problem 1: In $\mathbb{R}^4$, let $B := \{u_1, u_2, u_3, u_4\}$, where $u_1 := (4, 4, 1, 3)$, $u_2 := (3, 2, 2, 1)$, $u_3 := (1, 5, 1, 6)$, and $u_4 := (4, 2, 6, 7)$. Let $B' := \{v_1, v_2, v_3, v_4\}$, where $v_1 := (1, 2, 3, 4)$, $v_2 := (3, 1, 5, 5)$, $v_3 := (3, 1, 5, 1)$, and $v_4 := (2, 1, 1, 2)$. Assume that B and B' are ordered bases. Let w be the vector whose coordinate vector relative to B is $[w]_B := (4, 5, 6, 7)$.

1. Find w.

2. Find the transition matrix from B to B'.

3. Find the coordinate vector of w relative to B'.

Solution 1: First we enter and name the eight vectors as done in the first two lines of Figure 13.1. Since $[w]_B := (4, 5, 6, 7)$, then $w = 4u_1 + 5u_2 + 6u_3 + 7u_4$. Thus, we enter and evaluate as a matrix/vector the expression for w. The vector w is displayed in line 3 of Figure 13.1.

The transition matrix from B to B' is the matrix whose columns are $[u_1]_{B'}$, $[u_2]_{B'}$, $[u_3]_{B'}$, and $[u_4]_{B'}$. Thus, we need to find the B' coordinate vectors of u_1, u_2, u_3, and u_4.

To find $[u_1]_{B'}$ we need to solve the equation $u_1 = av_1 + bv_2 + cv_3 + dv_4$ for the variables a, b, c, d. Enter the equation u1 = a*v1 + b*v2 + c*v3 + d*v4; then **matsolve** to get the solution in line 5 of Figure 13.1. Put the values into the vector, **m1**. The vector **m1** makes up the first column of the transition matrix, which appears in line 1 of Figure 13.2.

Repeat this procedure for each of the others, $[u_2]_{B'}$, $[u_3]_{B'}$ and $[u_4]_{B'}$, to generate the vectors **m2**, **m3**, and **m4**. Define the transition matrix M by augmenting the four vectors m_i as in line 1 of Figure 13.2.

The B' coordinates of w can be found from $[w]_{B'} = M[w]_B$ where M is the transition matrix. Assuming that the transition matrix has been named M, we enter M &* vector([4,5,6,7]), which is seen in line 2. Evaluating this product as a matrix with **evalm**, we get the B' coordinate vector of w. Observe that in line 3 of Figure 13.2, we have checked the answer and it agrees with that calculated for w in line 3 of Figure 13.1.

```
> u1 := vector([4,4,1,3]):  u2 := vector([3,2,2,1]):
  u3 := vector([1,5,1,6]):  u4 := vector([4,2,6,7]):
> v1 := vector([1,2,3,4]):  v2 := vector([3,1,5,5]):
  v3 := vector([3,1,5,1]):  v4 := vector([2,1,1,2]):
> w := evalm(4*u1+5*u2+6*u3+7*u4);
```

$$w := \begin{bmatrix} 65 & 70 & 62 & 102 \end{bmatrix}$$

```
> eq := u1=a*v1+b*v2+c*v3+d*v4;
```

$$eq := u1 = a\,v1 + b\,v2 + c\,v3 + d\,v4$$

```
> matsolve(eq);
```

$$\left\{ c = \tfrac{1}{2},\; b = \tfrac{-3}{2},\; d = 3,\; a = 1 \right\}$$

```
> m1:=vector([1,-3/2,1/2,3]);
```

$$m1 := \begin{bmatrix} 1 & \tfrac{-3}{2} & \tfrac{1}{2} & 3 \end{bmatrix}$$

Figure 13.1: Finding a transition matrix

```
> M := augment(m1,m2,m3,m4);
```

$$M := \begin{bmatrix} 1 & \tfrac{4}{13} & \tfrac{32}{13} & \tfrac{4}{13} \\[4pt] \tfrac{-3}{2} & \tfrac{-10}{13} & \tfrac{-73}{13} & \tfrac{51}{52} \\[4pt] \tfrac{1}{2} & \tfrac{9}{13} & \tfrac{-11}{52} & \tfrac{-3}{52} \\[4pt] 3 & \tfrac{19}{13} & \tfrac{22}{13} & \tfrac{6}{13} \end{bmatrix}$$

```
> evalm(M &* vector([4,5,6,7]));
```

$$\begin{bmatrix} \tfrac{292}{13} & \tfrac{-593}{52} & \tfrac{197}{52} & \tfrac{425}{13} \end{bmatrix}$$

```
> evalm(292/13*v1 - 593/52*v2 + 197/52*v3 + 425/13*v4);
```

$$\begin{bmatrix} 65 & 70 & 62 & 102 \end{bmatrix}$$

Figure 13.2: Finding the coordinates of **w** relative to B'

<u>Solution 2</u>: The method we presented in Solution 1 is intended to show a step-by-step analysis, but it is tedious because we have to enter and solve four systems of equations and then put the results in a matrix. There is a more efficient way: One can show that if C is the matrix whose columns are the vectors $\mathbf{u}_1, \mathbf{u}_2, \mathbf{u}_3, \mathbf{u}_4$ and D is the matrix whose columns are the vectors $\mathbf{v}_1, \mathbf{v}_2, \mathbf{v}_3, \mathbf{v}_4$, then the four systems of equations we had to solve are equivalent to the single matrix equation $C = D A$. Its solution is $A = D^{-1}C$. (Explain why D must have an inverse.)
It is relatively efficient to enter the two matrices

$$C := \begin{bmatrix} 4 & 3 & 1 & 4 \\ 4 & 2 & 5 & 2 \\ 1 & 2 & 1 & 6 \\ 3 & 1 & 6 & 7 \end{bmatrix} \qquad D := \begin{bmatrix} 1 & 3 & 3 & 2 \\ 2 & 1 & 1 & 1 \\ 3 & 5 & 5 & 1 \\ 4 & 5 & 1 & 2 \end{bmatrix}$$

Next, evaluate as a matrix $A := D^{-1}C$. The result is line 3 of Figure 13.3. Observe that this agrees with the earlier calculation of M that appears in Figure 13.2.

```
> C := matrix(4,4,[4,3,1,4, 4,2,4,2, 1,2,1,6, 3,1,6,7]):
> D := matrix(4,4,[1,3,3,2, 2,1,1,1, 3,5,5,1, 4,5,1,2]):
> A := evalm(D^(-1) &* C);
```

$$A := \begin{bmatrix} 1 & \frac{4}{13} & \frac{32}{13} & \frac{4}{13} \\ \frac{-3}{2} & \frac{-10}{13} & \frac{-73}{13} & \frac{51}{52} \\ \frac{1}{2} & \frac{9}{13} & \frac{-11}{52} & \frac{-3}{52} \\ 3 & \frac{19}{13} & \frac{22}{13} & \frac{6}{13} \end{bmatrix}$$

Figure 13.3: | Alternative solution to Solved Problem 1 |

Solved Problem 2: Show that the matrix A below is orthogonal (a) by showing that $||A\mathbf{v}|| = ||\mathbf{v}||$ for all $\mathbf{v}$ and (b) by another method.

$$A := \begin{bmatrix} \frac{1}{2} & \frac{1}{2} & \frac{1}{2} & \frac{1}{2} \\ 0 & \frac{-1}{\sqrt{2}} & \frac{1}{\sqrt{2}} & 0 \\ \frac{-5}{6} & \frac{1}{6} & \frac{1}{6} & \frac{1}{2} \\ \frac{\sqrt{2}}{6} & \frac{-\sqrt{2}}{3} & \frac{-\sqrt{2}}{3} & \frac{1}{\sqrt{2}} \end{bmatrix}$$

<u>Solution</u>: If you find this matrix tedious to type, note that you can enter just the numerators and then use `mulrow` to divide each row by its denominator. We will assume that the reader has successfully entered the matrix and named it A.

We enter `n := evalm(A &* vector([x,y,z,w]))` as line 1 of Figure 13.4. Now use `innerprod(n,n)` for line 2, which we recognize as $||(x, y, z, w)||^2$. This is all we have to do to verify part (a); note that `n` shows the complexity of what we just did. The resulting expression is quite complicated. It is not at all obvious that the length of `n`, a huge vector, is the same as that of (x, y, z, w).

Now for part (b): The simplest way to verify that A is orthogonal is to show that $A A^t = I$ (number 2 in the theorem in the Introduction). Therefore, we enter `evalm(A &* transpose(A))` as line 3 of Figure 13.4. We see the result is the identity matrix and conclude that A is orthogonal.

```
> n:=evalm(A &* vector([x,y,z,w]));
```

$$n := \left[\tfrac{1}{2}x + \tfrac{1}{2}y + \tfrac{1}{2}z + \tfrac{1}{2}w \quad -\tfrac{1}{2}\sqrt{2}y + \tfrac{1}{2}\sqrt{2}z \quad -\tfrac{5}{6}x + \tfrac{1}{6}y + \tfrac{1}{6}z + \tfrac{1}{2}w \right.$$
$$\left. \tfrac{1}{6}\sqrt{2}x - \tfrac{1}{3}\sqrt{2}y - \tfrac{1}{3}\sqrt{2}z + \tfrac{1}{2}\sqrt{2}w \right]$$

```
> innerprod(n,n);
```

$$x^2 + y^2 + z^2 + w^2$$

```
> evalm(A &* transpose(A));
```

$$\begin{bmatrix} 1 & 0 & 0 & 0 \\ 0 & 1 & 0 & 0 \\ 0 & 0 & 1 & 0 \\ 0 & 0 & 0 & 1 \end{bmatrix}$$

Figure 13.4: | An orthogonal matrix preserves length |

13.3   Exercises

1. Assume that

$$B = \{(2,3,4,1),(4,4,5,2),(1,2,3,3),(1,1,1,1)\}$$
$$B' = \{(2,7,5,3),(1,4,2,2),(5,1,3,2),(7,7,4,5)\}$$

are two ordered bases for $\mathbb{R}^4$. Suppose that the B coordinate vector of $\mathbf{w}$ is $[\mathbf{w}]_B = (4,1,3,1)$.

(a) Find $\mathbf{w}$.

(b) Find the transition matrix from B to B'.

(c) Find the B' coordinate vector $[\mathbf{w}]_{B'}$ of $\mathbf{w}$.

(d) Find the transition matrix from B' to B.

(e) Multiply the matrices in (b) and (d). What do you conclude about them?

2. Let

$$B = \{x^3 - 2x^2 + 1, x^3 + x - 3, x^2 + 3x - 8, x^3 + 4x^2 - x - 1\}$$
$$B' = \{x^3 - 3x^2 + 2x, x^3 + x^2 + x + 1, x^3 + 7x - 3, x^3 + 4x^2 - 3x + 4\}$$

be two ordered bases for $\mathbb{P}_3$. Let $p(x)$ be the cubic polynomial whose B coordinate vector is $[p(x)]_B = (2,4,1,5)$.

(a) Find $p(x)$.

(b) Find the transition matrix from B to B'.

(c) Find the B' coordinate vector, $[p(x)]_{B'}$, of $p(x)$.

(d) Find the transition matrix from B' to B.

(e) Multiply the matrices in (b) and (d). What do you conclude about them?

3. Let

$$B = \left\{ \begin{bmatrix} 2 & 3 \\ 1 & 2 \end{bmatrix}, \begin{bmatrix} 3 & 6 \\ 1 & 1 \end{bmatrix}, \begin{bmatrix} 4 & 2 \\ 8 & 3 \end{bmatrix}, \begin{bmatrix} 1 & 3 \\ 4 & 2 \end{bmatrix} \right\}$$

$$B' = \left\{ \begin{bmatrix} 3 & 7 \\ 2 & 5 \end{bmatrix}, \begin{bmatrix} 4 & 5 \\ 2 & 2 \end{bmatrix}, \begin{bmatrix} 5 & 2 \\ 1 & 9 \end{bmatrix}, \begin{bmatrix} 3 & 3 \\ 2 & 6 \end{bmatrix} \right\}$$

be two ordered bases for $M_{2,2}$. Let $\mathbf{w}$ be the matrix whose B coordinate vector is $[\mathbf{w}]_B = (4, 1, 3, 3)$.

(a) Find $\mathbf{w}$.

(b) Find the transition matrix from B to B'.

(c) Find the B' coordinate vector, $[\mathbf{w}]_{B'}$, of $\mathbf{w}$.

(d) Find the transition matrix from B' to B.

(e) Multiply the matrices in (b) and (d). What do you conclude about them?

13.4   Exploration and Discovery

1. Use the examples of 4 by 4 orthogonal matrices below to test the following propositions. Provide counterexamples for those that are false, and prove the ones that are true.

$$
\begin{bmatrix}
\frac{1}{2} & \frac{-1}{2} & \frac{1}{2} & \frac{1}{2} \\[4pt]
\frac{-\sqrt{2}}{6} & \frac{1}{\sqrt{2}} & \frac{\sqrt{2}}{3} & \frac{\sqrt{2}}{3} \\[4pt]
\frac{5\sqrt{2}}{9} & \frac{\sqrt{2}}{3} & \frac{-5\sqrt{2}}{18} & \frac{\sqrt{2}}{18} \\[4pt]
\frac{-5}{18} & \frac{-1}{6} & \frac{-11}{18} & \frac{13}{18}
\end{bmatrix}
\qquad
\begin{bmatrix}
\frac{1}{\sqrt{3}} & 0 & \frac{1}{\sqrt{3}} & \frac{-1}{\sqrt{3}} \\[4pt]
\frac{-1}{\sqrt{3}} & \frac{1}{\sqrt{3}} & \frac{1}{\sqrt{3}} & 0 \\[4pt]
\frac{2}{\sqrt{15}} & \frac{1}{\sqrt{15}} & \frac{1}{\sqrt{15}} & \frac{1}{\sqrt{15}} \\[4pt]
\frac{-1}{\sqrt{15}} & \frac{-3}{\sqrt{15}} & \frac{2}{\sqrt{15}} & \frac{1}{\sqrt{15}}
\end{bmatrix}
$$

$$
\begin{bmatrix}
\frac{-1}{2} & \frac{1}{2} & \frac{1}{2} & \frac{-1}{2} \\[4pt]
0 & \frac{1}{\sqrt{6}} & \frac{1}{\sqrt{6}} & \frac{2}{\sqrt{6}} \\[4pt]
\frac{6}{\sqrt{66}} & \frac{-1}{\sqrt{66}} & \frac{5}{\sqrt{66}} & \frac{-2}{\sqrt{66}} \\[4pt]
\frac{-3\sqrt{11}}{22} & \frac{-5\sqrt{11}}{22} & \frac{3\sqrt{11}}{22} & \frac{\sqrt{11}}{22}
\end{bmatrix}
$$

(a) If A is orthogonal, then A^{-1} is orthogonal.

(b) If A and B are orthogonal, then AB is orthogonal.

(c) If A and B are orthogonal, then $A + B$ is orthogonal.

(d) If A is orthogonal, then $\det(A) = 1$.

2. Discuss how to generate examples of 5 by 5 orthogonal matrices (other than diagonal ones). Make up a few and check them. (Hint: You might make examples that have a lot of zeros. The Gram-Schmidt process may help you to find more interesting examples.)

13.5   Transition Matrices

LABORATORY EXERCISE 13–1

Name _____   Due Date _____

Suppose that

$$B = \{(1,2,3,1),(2,3,1,2),(1,1,2,3),(1,2,1,2)\}$$
$$B' = \{(2,1,1,3),(1,1,2,2),(2,1,3,3),(1,1,1,1)\}$$

are two bases for $\mathbb{R}^4$, and that the coordinate vector of $\mathbf{w}$ relative to B is $[\mathbf{w}]_B = (1,2,3,4)$.

1. Find $\mathbf{w}$.

2. Find the transition matrix from B to B'.

3. Find the B' coordinate vector $[\mathbf{w}]_{B'}$ of $\mathbf{w}$.

4. Find the transition matrix from B' to B.

5. Find the inverse of the matrix in 2. Compare the answer to that found in 4.

Chapter 14
EIGENVALUES AND EIGENVECTORS

LINEAR ALGEBRA CONCEPTS

- **Characteristic polynomial**
- **Eigenvalue**
- **Eigenvector**
- **Eigenspace**

14.1 Introduction

Maple's syntax for the characteristic polynomial of a matrix A is `charpoly(A,x)`. Maple can find the exact eigenvalues and eigenvectors of many matrices, but often both eigenvalues and eigenvectors must be approximated. The `linalg` package that comes with Maple contains the tools necessary to do this.

In many texts the Greek letter λ (lambda) is used for eigenvalues. Because Maple supports Greek characters, we will use λ. It will also be convenient for us to have a short name for the identity matrix; enter `alias(Id=&*())` to make `Id` an alias, or synomym, for I. (We don't use `I` because, to Maple, `I` is the complex number $\sqrt{-1}$.)

14.2 Solved Problems

Solved Problem 1: Let $A = \begin{bmatrix} 45 & -21 & -63 & -12 & 21 \\ 28 & -12 & -42 & -7 & 14 \\ 31 & -15 & -43 & -9 & 15 \\ 28 & -14 & -42 & -5 & 14 \\ 51 & -25 & -75 & -14 & 27 \end{bmatrix}$.

1. Find the characteristic polynomial of A.

2. Find the eigenvalues of A.

3. Find a basis for each eigenspace of A.

Solution: Enter the matrix and name it A. Note that we have typed the matrix over several lines to make it easier to enter.

To find the characteristic polynomial, we may use either `det(lambda*Id - A)` or `charpoly(A,lambda)`. The characteristic polynomial, p, appears in line 2 of Figure 14.1.

To find the eigenvalues, we **solve** the characteristic polynomial. The eigenvalues 2 and 3 appear in line 3 of Figure 14.1. Since 2 appears as a root 3 times, it is an eigenvalue of multiplicity 3; similarly, 3 is an eigenvalue of multiplicity 2.

```
> A := matrix(5,5,
     [45,-21,-63,-12,21,
     28,-12,-42,-7,14,
     31,-15,-43,-9,15,
     28,-14,-42,-5,14,
     51,-25,-75,-14,27]):
> p := charpoly(A, lambda);
```

$$p := \lambda^5 - 12\lambda^4 + 57\lambda^3 - 134\lambda^2 + 156\lambda - 72$$

```
> solve(p=0);
```

$$2, 2, 2, 3, 3$$

Figure 14.1: | Eigenvalues of a matrix |

The eigenspace corresponding to 2 is the nullspace of $2I - A$. Thus, we enter `rref(2*Id - A)`. The reduced matrix appears in line 1 of Figure 14.2. From this matrix, we obtain a basis for the nullspace (and hence for the eigenspace belonging to the eigenvalue 2) given below.

$$\begin{bmatrix} 0 \\ -3 \\ 1 \\ 0 \\ 0 \end{bmatrix}, \begin{bmatrix} \frac{3}{2} \\ \frac{5}{2} \\ 0 \\ 1 \\ 0 \end{bmatrix}, \begin{bmatrix} 0 \\ 1 \\ 0 \\ 0 \\ 1 \end{bmatrix}$$

In the same way, use rref(3*I-A) to get the basis for the nullspace displayed below.

$$\begin{bmatrix} -\frac{2}{7} \\ 1 \\ -\frac{5}{7} \\ 1 \\ 0 \end{bmatrix}, \quad \begin{bmatrix} 1 \\ 0 \\ 1 \\ 0 \\ 1 \end{bmatrix}$$

> rref(2*Id-A);

$$\begin{bmatrix} 1 & 0 & 0 & \frac{-3}{2} & 0 \\ 0 & 1 & 3 & \frac{-5}{2} & -1 \\ 0 & 0 & 0 & 0 & 0 \\ 0 & 0 & 0 & 0 & 0 \\ 0 & 0 & 0 & 0 & 0 \end{bmatrix}$$

> rref(3*Id-A);

$$\begin{bmatrix} 1 & 0 & 0 & \frac{2}{7} & -1 \\ 0 & 1 & 0 & -1 & 0 \\ 0 & 0 & 1 & \frac{5}{7} & -1 \\ 0 & 0 & 0 & 0 & 0 \\ 0 & 0 & 0 & 0 & 0 \end{bmatrix}$$

Figure 14.2: Eigenvectors of a matrix

Solved Problem 2: Find the eigenvalues and associated eigenvectors of

$$A = \begin{bmatrix} 3 & 2 & 5 \\ 3 & 4 & 1 \\ 6 & 7 & 3 \end{bmatrix}$$

Solution: The eigenvalues are found exactly as in Solved Problem 1. One of the eigenvalues is the extremely complicated line 3 of Figure 14.3. When we seek its corresponding eigenvectors, serious problems occur as we will see next when we try to row-reduce $wI - A$.

Try to calculate rref(eigvals[1]*Id-A)). We can get a decimal approximation to the eigenvalue by using evalf. Note the very small imaginary part; it can be removed by applying fnormal, Maple's normalizing function that helps eliminate roundoff errors in decimal calculations. (Try using Maple's simplifying functions, `simplify`, `normal`, and `combine`, to eliminate the imaginary part from `eigvals[1]`. A graph of the characteristic polynomial shows the imaginary part is actually zero.) If we try to get around this difficulty by using a decimal approximation, then Maple produces the 3 by 3 identity matrix in line 4 of Figure 14.4. This is incorrect, of course, because an eigenvalue is a number w that makes the matrix $wI - A$ singular; it cannot reduce to the identity. What's going on?

If Maple were to try to row-reduce the matrix $wI - A$, where w is as in line 3 of Figure 14.4, it would encounter entries that are actually equivalent to zero, but are so complicated that Maple can't recognize them. At some point, an unrecognized division by zero occurs, and an incorrect reduced matrix is returned. Recognizing complicated expressions as zero is one of the most difficult problems for computer mathematics systems and is a current research topic. Thus, we will settle for approximate eigenvectors.

```
> A := matrix(3,3,[3,2,5, 3,4,1, 6,7,3]);
```

$$A := \begin{bmatrix} 3 & 2 & 5 \\ 3 & 4 & 1 \\ 6 & 7 & 3 \end{bmatrix}$$

```
> charpoly(A,lambda);
```

$$\lambda^3 - 10\lambda^2 - 10\lambda + 6$$

```
> eigvals := [solve(")];
```

$$eigvals := \left[\%1^{1/3} + \tfrac{130}{9}\tfrac{1}{\%1^{1/3}} + \tfrac{10}{3}, \right.$$
$$-\tfrac{1}{2}\%1^{1/3} - \tfrac{65}{9}\tfrac{1}{\%1^{1/3}} + \tfrac{10}{3} + \tfrac{1}{2}I\sqrt{3}\left(\%1^{1/3} - \tfrac{130}{9}\tfrac{1}{\%1^{1/3}}\right)$$
$$\left. -\tfrac{1}{2}\%1^{1/3} - \tfrac{65}{9}\tfrac{1}{\%1^{1/3}} + \tfrac{10}{3} - \tfrac{1}{2}I\sqrt{3}\left(\%1^{1/3} - \tfrac{130}{9}\tfrac{1}{\%1^{1/3}}\right)\right]$$
$$\%1 := \tfrac{1369}{27} + \tfrac{1}{9}I\sqrt{35871}$$

Figure 14.3: A matrix with complicated eigenvalues

Since the problem occurs because of arithmetic with complicated numbers, we will try to reduce the matrix *before* substituting the eigenvalues. Execute

```
ffgausselim(x*Id - A)
```

```
> rref(eigvals[1]*Id - A);
Error, (in rref)
matrix entries must be rational polynomials
> evalf(eigvals[1]);
```
$$10.86924045 + .1\,10^{-9}I$$

```
> w := fnormal(");
```
$$w := 10.86924045$$

```
> rref(w*Id - A);
```
$$\begin{bmatrix} 1 & 0 & 0 \\ 0 & 1 & 0 \\ 0 & 0 & 1 \end{bmatrix}$$

```
> M := ffgausselim(x*Id - A);
```
$$\begin{bmatrix} -3 & x-4 & -1 \\ 0 & -3+6x & -3x+3 \\ 0 & 0 & x^3 - 10x^2 - 10x + 6 \end{bmatrix}$$

```
> M[3,3] := 0:
> subs(x=w, evalm(M));
```
$$\begin{bmatrix} -3 & 6.86924045 & -1 \\ 0 & 62.21544270 & -29.60772135 \\ 0 & 0 & 0 \end{bmatrix}$$

```
> nullspace("); v := "[1]:
```
$$\{[.7563348063 \quad .4758902302 \quad 1]\}$$

```
> evalm(A &* v = w*v);
```
$$[8.220784879 \quad 5.172565340 \quad 10.86924045] =$$
$$[8.220784870 \quad 5.172565340 \quad 10.86924045]$$

Figure 14.4: Approximating eigenvectors

naming it M as in Figure 14.4. The name **ffgausselim** stands for **fraction-free gausselimination**. We recognize the entry $M_{3,3}$ as the characteristic polynomial of A. When we substitute any eigenvalue, this term must be zero; make it so with **M[3,3] := 0**. We are ready to calculate approximate eigenvectors. Substitute w for x in M. A basis for the nullspace of M, and hence the eigenspace for A, can be read from the result or **nullspace** can be applied.

The approximate eigenvector is in line 8 in Figure 14.4.

What does "approximate" mean? Just how close is our result to a true eigenvector? Maple gives no way to tell, but the following idea gives us some "evidence" of nearness:

> If w is a true eigenvalue and $\mathbf{v}$ a corresponding true eigenvector, then $A\mathbf{v} = w\mathbf{v}$. Therefore, if the eigenvalue and corresponding eigenvector are good approximations, the two sides of the equation should be very close to each other.

Let's test this: calculate **A &* v = w*v** as in line 9 in Figure 14.4. A times the approximate eigenvector appears in the first row of line 16, and 10.86924045 times the approximate eigenvector appears in the second row. These two vectors agree through eight decimal places, so we are led to suspect that (.7563348063, .4758902302, 1) is very near to some true eigenvector belonging to the exact eigenvalue in line 5 of Figure 14.3. But this is just evidence. We cannot conclude anything quantitatively about *how* near. It depends on A and we have no way to estimate it.

We leave it to the reader to approximate and check the eigenvectors for the remaining two eigenvalues in a similar way.

Solved Problem 3: Approximate the eigenvalues and eigenvectors of

$$A := \begin{bmatrix} 1 & 1 & 2 & 1 & 1 \\ 1 & 2 & 3 & 2 & 1 \\ 2 & 3 & 1 & 2 & 1 \\ 1 & 2 & 2 & 3 & 1 \\ 1 & 1 & 1 & 1 & 7 \end{bmatrix}$$

Solution: Enter the 5 by 5 matrix and name it A. Next, evaluate **charpoly(A, lambda)**. The characteristic polynomial appears in line 2 of Figure 14.5. If we ask Maple to solve this, it will fail to find exact values. Thus, we must be satisfied with an approximation of the zeros of the characteristic polynomial. Obtain the approximate roots with **fsolve**.

```
> A := matrix(5,5,[1,1,2,1,1, 1,2,3,2,1, 2,3,1,2,1, 1,2,2,3,1,
    1,1,1,1,7]):
> pa := charpoly(A, lambda);
```

$$pa := \lambda^5 - 14\,\lambda^4 + 39\,\lambda^3 + 59\,\lambda^2 - 121\,\lambda + 31;$$

```
> solve(pa=0);
```

$$\text{RootOf}\left(\_Z^5 - 14\,\_Z^4 + 39\,\_Z^3 + 59\,\_Z^2 - 121\,\_Z + 31\right)$$

```
> fsolve(pa=0);
```

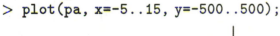

$$-1.898437230 \quad .3126193022 \quad 1.071200946 \quad 5.280735608 \quad 9.233881374$$

Figure 14.5: | Approximating eigenvalues of a matrix |

To visually check that we have all the roots and that the values are (approximately) correct, plot the characteristic polynomial over a suitable interval.

Finally, we ask Maple to solve on each of the intervals listed above. See sections A.7 and A.11 in Appendix A for details on solving equations. The approximate eigenvalues appear in line 4 of Figure 14.5. The eigenvectors should be approximated and checked for accuracy using the the same technique as in Solved Problem 2.

```
> plot(pa, x=-5..15, y=-500..500);
```

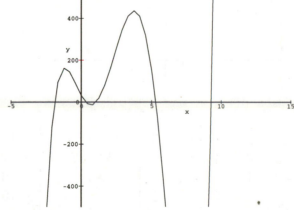

Figure 14.6: | The graph of a characteristic polynomial |

14.3   Exercises

1. For each of the following matrices, find the characteristic polynomial, the exact eigenvalues, and a basis for each eigenspace, exactly.

$$A := \begin{bmatrix} 3 & 4 & -4 & -4 \\ 4 & 3 & -4 & -4 \\ 0 & 4 & -1 & -4 \\ 4 & 0 & -4 & -1 \end{bmatrix} \qquad B := \begin{bmatrix} 0 & 4 & 4 & -4 \\ -6 & 10 & 6 & -6 \\ -2 & 2 & 6 & -2 \\ -6 & 6 & 6 & -2 \end{bmatrix}$$

2. For each of the following matrices:

   (a) Find the characteristic polynomial.

   (b) Approximate the eigenvalues and report their exact values where it is practical to do so, that is, if they can be found exactly and are not immensely complicated.

   (c) Find a basis for each eigenspace. Where it is practical to do so, report the eigenvectors exactly. Otherwise approximate them.

$$A = \begin{bmatrix} -2 & 4 & 4 & 3 \\ -2 & 9 & 3 & 0 \\ -1 & 4 & 2 & 1 \\ 1 & 4 & 2 & 1 \end{bmatrix} \qquad B = \begin{bmatrix} -5 & 11 & 0 & -1 \\ -4 & 12 & 1 & 0 \\ -4 & 9 & 0 & -2 \\ -2 & 8 & 0 & 1 \end{bmatrix}$$

$$C = \begin{bmatrix} -2 & 5 & 4 & 0 \\ -3 & 9 & 3 & -1 \\ -1 & 3 & 4 & 0 \\ -2 & 7 & 1 & -1 \end{bmatrix} \qquad D = \begin{bmatrix} -2 & -4 & 8 & -2 & 8 \\ -16 & 2 & 18 & 6 & 0 \\ -2 & -4 & 8 & -3 & 8 \\ 8 & -6 & -4 & -6 & 12 \\ -2 & -4 & 6 & -3 & 10 \end{bmatrix}$$

$$E = \begin{bmatrix} -3 & 8 & 0 & 8 & -1 \\ 0 & 12 & -4 & 10 & -3 \\ -3 & 8 & 1 & 9 & 1 \\ -2 & 1 & 3 & 3 & 2 \\ -4 & 7 & 3 & 6 & -1 \end{bmatrix} \qquad F = \begin{bmatrix} -9 & 2 & 7 & 2 & 5 \\ -11 & 0 & 9 & 0 & 7 \\ -21 & 2 & 17 & 1 & 8 \\ -15 & 2 & 12 & 1 & 3 \\ -10 & 1 & 10 & -1 & 4 \end{bmatrix}$$

3. For each of the following matrices, find the characteristic polynomial. Approximate the eigenvalues and approximate the eigenvectors forming a basis for the

eigenspace corresponding to each eigenvalue.

$$A := \begin{bmatrix} 12 & 0 & -7 & -7 & 7 \\ 10 & 1 & -5 & -6 & 6 \\ 7 & 1 & -4 & -4 & 4 \\ 12 & -1 & -7 & -7 & 6 \\ 8 & 1 & -4 & -4 & 3 \end{bmatrix} \qquad B := \begin{bmatrix} 4 & 3 & -1 & -4 & 5 \\ 0 & 5 & 1 & 0 & 3 \\ 7 & 1 & -3 & -5 & 5 \\ 2 & 3 & 0 & -3 & 3 \\ 3 & 3 & 0 & -2 & 2 \end{bmatrix}$$

$$C := \begin{bmatrix} 21 & -1 & -10 & -13 & 12 \\ 18 & -3 & -5 & -11 & 11 \\ 13 & 2 & -6 & -9 & 8 \\ 17 & -1 & -9 & -10 & 9 \\ 19 & -1 & -9 & -12 & 10 \end{bmatrix}$$

4. Assume that A is a 4 by 4 matrix with eigenvalues 2, 3, and 4. Bases for the corresponding eigenspaces are given in the following table. Find A.

Eigenvalue	4	3	2
Basis for eigenspace	(3,1,2,1)	(4,4,2,3)	(2,1,1,1), (3,2,2,1)

Hint: Let A be a matrix with variable entries, and look at the following system of equations.

$$A \begin{bmatrix} 2 \\ 1 \\ 1 \\ 1 \end{bmatrix} = 2 \begin{bmatrix} 2 \\ 1 \\ 1 \\ 1 \end{bmatrix}, \qquad A \begin{bmatrix} 3 \\ 2 \\ 2 \\ 1 \end{bmatrix} = 2 \begin{bmatrix} 3 \\ 2 \\ 2 \\ 1 \end{bmatrix},$$

$$A \begin{bmatrix} 4 \\ 4 \\ 2 \\ 3 \end{bmatrix} = 3 \begin{bmatrix} 4 \\ 4 \\ 2 \\ 3 \end{bmatrix}, \qquad A \begin{bmatrix} 3 \\ 1 \\ 2 \\ 1 \end{bmatrix} = 4 \begin{bmatrix} 3 \\ 1 \\ 2 \\ 1 \end{bmatrix}.$$

14.4   Exploration and Discovery

1. We will explore the eigenvalues of functions of the matrix A of Exercise 1.

   (a) Find the eigenvalues of the matrix A of Exercise 1.

   (b) Find the eigenvalues of A^2 and A^3.

   (c) Find the eigenvalues of $A^2 + A^3$.

   (d) Find the eigenvalues of $p(A)$, where $p(x) = 3x^3 - 2x^2 + 1$.

   (e) Explain how your answers to (b) through (d) relate to your answer in (a). Formulate a general conjecture and try to prove it.

2. **The Cayley-Hamilton Theorem**. For each of the three matrices in Exercise 1, calculate the characteristic polynomial $p(x)$ and evaluate $p(x)$ at the matrix. (See Chapter 3 to recall what this means.) Do Exploration and Discovery Problem 2 in Chapter 5.

3. Calculate the characteristic polynomials of A, B, and C in Exercise 2 and of their transposes. Discuss your observations and conclusions.

4. If A denotes the matrix from Exercise 2 and $\mathbf{v} = [1, 1, 1, 1]$, approx $A^5\mathbf{v}$, $A^{10}\mathbf{v}$, $A^{50}\mathbf{v}$, and $A^{100}\mathbf{v}$. Compare the vectors you obtain with the eigenvectors of A. Repeat the experiment using different (nonzero) vectors $\mathbf{v}$ and matrices B and C from Exercise 2. Report your findings.

5. The *trace* of a square matrix is defined to be the sum of its diagonal entries. Thus, the trace of $\begin{bmatrix} -14 & -24 & -19 \\ 1 & 5 & 1 \\ 14 & 18 & 19 \end{bmatrix}$ is 10.

   (a) For the matrix above, compute the trace and the characteristic polynomial. Make up a 4 by 4, a 5 by 5, and a new 3 by 3 matrix and do the same.

   (b) Compare the characteristic polynomials and the traces above. What do you notice?

   (c) Enter a 2 by 2 and a 3 by 3 matrix, all of whose entries are variables. Compare their characteristic polynomials and traces. What do you notice?

(d) Use the `solve` command to find the roots of the two characteristic polynomials of the 3 by 3 examples you used in (a) and (b). Find the sum of these roots. What do you notice?

(e) Prove that the coefficient of x^{n-1} in a polynomial $x^n + a_{n-1}x^{n-1} + \ldots + a_1 x + a_0$ is the sum of the roots of the polynomial. (<u>Hint</u>: Every such polynomial factors as $(x-r_1)(x-r_2)\ldots(x-r_n)$. The r_i's may be complex or may be repeated. Use Maple to `expand` such products.)

(f) Summarize your observations and conclusions from all the above. What do your observations say about the relationship between the eigenvalues and the trace of a matrix?

(g) Show that if A is similar to B then they have the same trace. (<u>Hint</u>: Show that similar matrices have the same determinant. Show that $A - wI$ and $B - wI$ are similar and hence A and B have the same characteristic polynomial.)

14.5   Eigenvalues and Eigenvectors

LABORATORY EXERCISE 14–1

Name _____   **Due Date** _____

Let $A := \begin{bmatrix} 1 & -2 & 4 & 2 \\ -2 & 1 & 4 & 2 \\ 0 & -2 & 5 & 2 \\ -2 & -2 & 4 & 5 \end{bmatrix}$.

1. Find the characteristic polynomial of A.

2. Find the exact eigenvalues of A.

3. Find a basis for each eigenspace of A.

Chapter 15
DIAGONALIZATION AND ORTHOGONAL DIAGONALIZATION

LINEAR ALGEBRA CONCEPTS

- **Similarity**
- **Diagonalization**
- **Symmetric matrix**
- **Orthogonal diagonalization**

15.1   Introduction

Two matrices A and B are called *similar* if there is some invertible matrix P such that $B = P^{-1}AP$.

An n by n matrix is called *diagonalizable* if it is similar to a diagonal matrix. This is the case if it has a set of eigenvectors that forms a basis for $\mathbb{R}^n$. In many instances, a solution to a problem for diagonal matrices can be carried over to diagonalizable matrices by similarity.

15.2   Solved Problems

Solved Problem 1: Determine if the following matrices are diagonalizable. If they are, exhibit the diagonalization.

$$A := \begin{bmatrix} 6 & 4 & -6 & 0 & 2 \\ 2 & 5 & -4 & 0 & 2 \\ 4 & 5 & -5 & 0 & 3 \\ 2 & 3 & -4 & 2 & 2 \\ 2 & 3 & -4 & -2 & 6 \end{bmatrix} \qquad B := \begin{bmatrix} -6 & -5 & 12 & -9 & 2 \\ -4 & 1 & 5 & -2 & -1 \\ -8 & -3 & 13 & -7 & 0 \\ -4 & -1 & 5 & 0 & -1 \\ -4 & -1 & 5 & -1 & 0 \end{bmatrix}$$

Solution for A: To make the worksheet more readable, use `alias(Id=&*())` to define `Id` as a synonym for the identity matrix. Enter the first 5 by 5 matrix and name it A. To find the eigenvalues use `charpoly(A,lambda)`. Now `solve` the characteristic polynomial. Maple's response shows not only the eigenvalues but their orders as well.

We could have also `factored` to obtain the same information. From line 2 of Figure 15.1, we see that 4 is an eigenvalue of order 1, and that 3 and 2 are eigenvalues of order 2.

```
> charpoly(A,lambda);
```

$$\lambda^5 - 14\,\lambda^4 + 77\,\lambda^3 - 208\,\lambda^2 + 276\,\lambda - 144$$

```
> solve(");
```

$$4, 2, 2, 3, 3$$

Figure 15.1: | Characteristic polynomial and eigenvalues of A |

In order to find the associated eigenvectors, row-reduce with the respective eigenvalues, `rref(4*Id-A)`, `rref(3*Id-A)`, and `rref(2*Id-A)`, as separate expressions. From lines 1, 2, and 3 of Figure 15.2 we deduce the following:

$\left\{ \left(\frac{2}{3}, \frac{2}{3}, 1, \frac{2}{3}, 1 \right) \right\}$ is a basis of the eigenspace for the eigenvalue 4.
$\{ (2, 0, 1, 0, 0), \; (-2, 1, 0, 1, 1) \}$ is a basis of the eigenspace for the eigenvalue 3.
$\left\{ \left(\frac{1}{2}, 1, 1, 0, 0 \right), \; \left(\frac{1}{2}, -1, 0, 1, 1 \right) \right\}$ is a basis of the eigenspace for the eigenvalue 2.

We have found five linearly independent eigenvectors for the 5 by 5 matrix A; therefore, we conclude that A is diagonalizable. We use this information to exhibit the diagonalization as follows:

$$\text{If } D := \begin{bmatrix} 4 & 0 & 0 & 0 & 0 \\ 0 & 3 & 0 & 0 & 0 \\ 0 & 0 & 3 & 0 & 0 \\ 0 & 0 & 0 & 2 & 0 \\ 0 & 0 & 0 & 0 & 2 \end{bmatrix} \text{ and } P := \begin{bmatrix} \frac{2}{3} & 2 & -2 & \frac{1}{2} & \frac{1}{2} \\ \frac{2}{3} & 0 & 1 & 1 & -1 \\ 1 & 1 & 0 & 1 & 0 \\ \frac{2}{3} & 0 & 1 & 0 & 1 \\ 1 & 0 & 1 & 0 & 1 \end{bmatrix}, \text{ then } P^{-1}AP = D.$$

We remark that our choice of P and D are not the only correct ones. The eigenvalues down the diagonal of D may be put in a different order and the columns of P interchanged accordingly. Also, eigenspaces are unique, but there are infinitely many bases for each eigenspace; thus, there are infinitely many correct matrices P.

To be certain that our diagonalization is correct, we need only check with Maple that $P^{-1}AP = D$. *When diagonalizing matrices, you should always perform this check.*

> `rref(4*Id-A);`

$$\begin{bmatrix} 1 & 0 & 0 & 0 & \frac{-2}{3} \\ 0 & 1 & 0 & 0 & \frac{-2}{3} \\ 0 & 0 & 1 & 0 & -1 \\ 0 & 0 & 0 & 1 & \frac{-2}{3} \\ 0 & 0 & 0 & 0 & 0 \end{bmatrix}$$

> `rref(3*Id-A);`

$$\begin{bmatrix} 1 & 0 & -2 & 0 & 2 \\ 0 & 1 & 0 & 0 & -1 \\ 0 & 0 & 0 & 1 & -1 \\ 0 & 0 & 0 & 0 & 0 \\ 0 & 0 & 0 & 0 & 0 \end{bmatrix}$$

> `rref(2*Id-A);`

$$\begin{bmatrix} 1 & 0 & \frac{-1}{2} & 0 & \frac{-1}{2} \\ 0 & 1 & -1 & 0 & 1 \\ 0 & 0 & 0 & 1 & -1 \\ 0 & 0 & 0 & 0 & 0 \\ 0 & 0 & 0 & 0 & 0 \end{bmatrix}$$

Figure 15.2: | Eigenvectors of A |

Solution for B: Enter and name the matrix B. Now `factor` the `charpoly(B, lambda)` as done in line 1 of Figure 15.3. We focus attention on the eigenvalue 2 of order 2. Row-reduce $2I - B$ with `rref(2*Id-B)`. We see from line 2 of Figure 15.3 that the nullspace of this matrix has dimension 1, and hence the eigenspace of B associated with the eigenvalue 2 has dimension 1. Since the eigenvalue 2 has order 2, we conclude without further calculation that B is not a diagonalizable matrix.

```
> factor(charpoly(B, lambda));
```

$$\lambda\,(\lambda-1)\,(\lambda-3)\,(\lambda-2)^2$$

```
> rref(2*Id - B);
```

$$\begin{bmatrix} 1 & 0 & 0 & 0 & \frac{-3}{2} \\ 0 & 1 & 0 & 0 & -1 \\ 0 & 0 & 1 & 0 & -2 \\ 0 & 0 & 0 & 1 & -1 \\ 0 & 0 & 0 & 0 & 0 \end{bmatrix}$$

Figure 15.3: $\boxed{\text{Eigenvalues and eigenvectors of } B}$

Solved Problem 2: Let $A := \begin{bmatrix} 0 & -\frac{1}{6} & -\frac{1}{6} & -\frac{1}{2} \\ -4 & -\frac{1}{3} & -\frac{4}{3} & -4 \\ -4 & -\frac{4}{3} & -\frac{1}{3} & -4 \\ 3 & \frac{5}{6} & \frac{5}{6} & \frac{7}{2} \end{bmatrix}$.

1. Find A^n for an arbitrary positive integer n.

2. $\boxed{\textbf{For students who have studied calculus.}}$ Find $\lim\limits_{n\to\infty} A^n$.

3. Find a cube root of A with real entries.

Solution: The first step is to diagonalize the matrix exactly as we did in Solved Problem 1. We conclude that

$$D = P^{-1}AP, \text{ where } P := \begin{bmatrix} -\frac{1}{2} & -1 & 0 & 0 \\ -1 & 0 & -1 & -3 \\ -1 & 0 & 1 & 0 \\ 1 & 1 & 0 & 1 \end{bmatrix} \text{ and } D := \begin{bmatrix} \frac{1}{3} & 0 & 0 & 0 \\ 0 & \frac{1}{2} & 0 & 0 \\ 0 & 0 & 1 & 0 \\ 0 & 0 & 0 & 1 \end{bmatrix}.$$

To calculate A^n we use the fact that $A^n = (PDP^{-1})^n = PD^nP^{-1}$. (You should be able to prove this, directly for $n = 2$, and then by induction for all $n > 2$.) Thus, with P and D appropriately defined as P and M (with the diagonal matrix command diag), we evaluate as a matrix P &* M^n &* P^(-1). (We are using M for D since D is Maple's differential operator and the name is protected.) The nth power of A is displayed as A_n in line 2 of Figure 15.4.

> M := diag(3^(-n), 2^(-n), 1, 1);

$$M := \begin{bmatrix} 3^{(-n)} & 0 & 0 & 0 \\ 0 & 2^{(-n)} & 0 & 0 \\ 0 & 0 & 1 & 0 \\ 0 & 0 & 0 & 1 \end{bmatrix}$$

> A_n := evalm(P &* M &* P^(-1));

$$A\_n :=$$
$$\begin{bmatrix} 3\cdot3^{(-n)} - 2\cdot2^{(-n)} & 3^{(-n)} - 2^{(-n)} & 3^{(-n)} - 2^{(-n)} & 3\cdot3^{(-n)} - 2\cdot2^{(-n)} \\ 6\cdot3^{(-n)} - 6 & 2\cdot3^{(-n)} - 1 & 2\cdot3^{(-n)} - 2 & 6\cdot3^{(-n)} - 6 \\ 6\cdot3^{(-n)} - 6 & 2\cdot3^{(-n)} - 2 & 2\cdot3^{(-n)} - 1 & 6\cdot3^{(-n)} - 6 \\ \dfrac{-6}{3^n} + \dfrac{2}{2^n} + 4 & -2\cdot3^{(-n)} + 2^{(-n)} + 1 & -2\cdot3^{(-n)} + 2^{(-n)} + 1 & \dfrac{-6}{3^n} + \dfrac{3}{2^n} + 4 \end{bmatrix}$$

Figure 15.4: The nth power of a matrix

To answer 2, we may use limit. To apply limit to all the entries of A_n, use map. The general syntax is map(*function*, *object*, *parameters*) to apply *function* to each element of *object* where *parameters* are extra arguments needed by *function*. The result of taking the limit of A_n as n goes to infinity is in Figure 15.5.

> map(limit, A_n, n = infinity);

$$\begin{bmatrix} 0 & 0 & 0 & 0 \\ -6 & -1 & -2 & -6 \\ -6 & -2 & -1 & -6 \\ 4 & 1 & 1 & 4 \end{bmatrix}$$

Figure 15.5: $\displaystyle\lim_{n\to\infty} A^n$

Alternative solution to 2: Observe that

$$\lim_{n\to\infty} A^n = P(\lim_{n\to\infty} D^n)P^{-1} = P \begin{bmatrix} 0 & 0 & 0 & 0 \\ 0 & 0 & 0 & 0 \\ 0 & 0 & 1 & 0 \\ 0 & 0 & 0 & 1 \end{bmatrix} P^{-1}$$

Use `Q := diag(0,0,1,1)` to enter, and name Q, the diagonal matrix shown above. Calculate `P &* Q &* P^(-1)` and the result will be the same as before.

To answer 3, we seek a matrix C with real entries such that $C^3 = A$. Notice that

$$(PD^{\frac{1}{3}}P^{-1})^3 = P(D^{\frac{1}{3}})^3 P^{-1} = PDP^{-1} = A$$

Thus $A^{\frac{1}{3}} = PD^{\frac{1}{3}}P^{-1}$. However, Maple will not calculate roots of matrices. It may occur to you to substitute $\frac{1}{3}$ for n in line 1 of Figure 15.4. This works, but it has to be justified because the formula was only shown to be valid for positive integers. The argument above presents the justification. Figure 15.6 shows the result of setting $n = \frac{1}{3}$. (We will not do it here, but you should check this answer by cubing it. Does it give the original matrix?)

```
> subs(n=1/3, evalm(A_n));
```

$$\begin{bmatrix} 3^{2/3} - 2^{2/3} & \frac{1}{3}3^{2/3} - \frac{1}{2}2^{2/3} & \frac{1}{3}3^{2/3} - \frac{1}{2}2^{2/3} & 3^{2/3} - \frac{3}{2}2^{2/3} \\ 2\,3^{2/3} - 6 & \frac{2}{3}3^{2/3} - 1 & \frac{2}{3}3^{2/3} - 2 & 2\,3^{2/3} - 6 \\ 2\,3^{2/3} - 6 & \frac{2}{3}3^{2/3} - 2 & \frac{2}{3}3^{2/3} - 1 & 2\,3^{2/3} - 6 \\ -2\,3^{2/3} + 2^{2/3} + 4 & -\frac{2}{3}3^{2/3} + \frac{1}{2}2^{2/3} + 1 & -\frac{2}{3}3^{2/3} + \frac{1}{2}2^{2/3} + 1 & -2\,3^{2/3} + \frac{3}{2}2^{2/3} + 4 \end{bmatrix}$$

Figure 15.6: $\boxed{\text{A cube root of a matrix}}$

Solved Problem 3: Orthogonally diagonalize the symmetric matrix

$$A := \begin{bmatrix} 3 & -1 & -1 & 1 \\ -1 & 3 & 1 & -1 \\ -1 & 1 & 3 & -1 \\ 1 & -1 & -1 & 3 \end{bmatrix}$$

Also, find a matrix B such that $B B^t = A$.

<u>Solution:</u> To orthogonally diagonalize a symmetric matrix, we have to find an *orthogonal* matrix Q such that $Q^{-1}AQ$ is diagonal. (Recall that Q is orthogonal, when $Q^{-1} = Q^t$.) Enter the matrix A and follow the procedure in Solved Problem 1 to find the eigenvalues and eigenvectors for A. From lines 2 and 3 of Figure 15.7, we obtain the following:

$\{(1, -1, -1, 1)\}$ is a basis of the eigenspace for the eigenvalue 6.

$\{(1, 1, 0, 0), (1, 0, 1, 0), (-1, 0, 0, 1)\}$ is a basis of the eigenspace for the eigenvalue 2.

<u>Remark:</u> Notice that the basis vector for the eigenvalue 6 is orthogonal to all the basis vectors for eigenvalue 2. When A is symmetric, vectors in eigenspaces corresponding to distinct eigenvalues are orthogonal.

```
> factor(charpoly(A,lambda));
```

$$(\lambda - 6)(\lambda - 2)^3$$

```
> rref(6*Id-A);
```

$$\begin{bmatrix} 1 & 0 & 0 & -1 \\ 0 & 1 & 0 & 1 \\ 0 & 0 & 1 & 1 \\ 0 & 0 & 0 & 0 \end{bmatrix}$$

```
> rref(2*Id-A);
```

$$\begin{bmatrix} 1 & -1 & -1 & 1 \\ 0 & 0 & 0 & 0 \\ 0 & 0 & 0 & 0 \\ 0 & 0 & 0 & 0 \end{bmatrix}$$

Figure 15.7: | Eigenvalues and eigenvectors for a symmetric matrix |

Now we have to convert these bases into orthonormal bases that will then form the columns of the matrix Q. The eigenspace for the eigenvalue 6 has dimension 1; thus, we find an orthonormal basis by simply normalizing (by hand) the single basis vector $(1, -1, -1, 1)$: set v0 equal $(\frac{1}{2}, -\frac{1}{2}, -\frac{1}{2}, \frac{1}{2})$. The orthonormal basis for the eigenspace of 6 appears in line 2 of Figure 15.8. The eigenspace for the eigenvalue 2 has dimension

3, and hence, we must apply the Gram-Schmidt process to obtain an orthonormal basis. This is seen in lines 2 through 6 of Figure 15.8. Recall, the 2-norm is the usual Euclidean length. To help with the process, define the projection function `proj` by

$$\texttt{proj := (u,v) -> dotprod(u,v)/dotprod(v,v) * v}$$

```
> proj := (u,v) -> dotprod(u,v)/dotprod(v,v) * v:
> v0:=vector([1/2,-1/2,-1/2,1/2]):
> u1 := vector([1,1,0,0]):
  u2 := vector([1,0,1,0]):
  u3 := vector([-1,0,0,1]):
> v1 := evalm(u1/norm(u1,2));
```

$$v1 := \left[\frac{1}{2}\sqrt{2} \quad \frac{1}{2}\sqrt{2} \quad 0 \quad 0 \right]$$

```
> u2 - proj(u2,v1);
  v2 := evalm(''/norm('',2));
```

$$u2 - \tfrac{1}{2}\, v1\, \sqrt{2}$$
$$v2 := \left[\tfrac{1}{6}\sqrt{6} \quad -\tfrac{1}{6}\sqrt{6} \quad \tfrac{1}{3}\sqrt{6} \quad 0 \right]$$

```
> u3 - proj(u3,v1)- proj(u3,v2);
  v3 := evalm(''/norm('',2));
```

$$u3 + \tfrac{1}{2}\, v1\, \sqrt{2} + \tfrac{1}{6}\, v2\, \sqrt{6}$$
$$v3 := \left[-\tfrac{1}{6}\sqrt{3} \quad \tfrac{1}{6}\sqrt{3} \quad \tfrac{1}{6}\sqrt{3} \quad \tfrac{1}{2}\sqrt{3} \right]$$

Figure 15.8: Finding an orthonormal basis

The desired matrix, Q, should have as its columns the vector v0 and the normalized vectors v1, v2, and v3. This is accomplished when we enter

$$\texttt{Q := augment(v0,v1,v2,v3);}$$

The result is seen as line 1 in Figure 15.9.

We will not do so here, but you should use Maple to check that $Q^t Q = I$ and that

$Q^t A Q$ is the diagonal matrix $D :=$
$\begin{bmatrix} 6 & 0 & 0 & 0 \\ 0 & 2 & 0 & 0 \\ 0 & 0 & 2 & 0 \\ 0 & 0 & 0 & 2 \end{bmatrix}$.

To solve the second part of the problem, we observe first that if

$$E := \begin{bmatrix} \sqrt{6} & 0 & 0 & 0 \\ 0 & \sqrt{2} & 0 & 0 \\ 0 & 0 & \sqrt{2} & 0 \\ 0 & 0 & 0 & \sqrt{2} \end{bmatrix},$$

then $EE^t = E^2 = D$. Thus $A = QDQ^t = Q(EE^t)Q^t = (QE)(QE)^t$. The solution $B = QE$ is line 2 of Figure 15.9. (Once again, you should use Maple to check that $B^t B = A$.)

```
> Q := augment(v0,v1,v2,v3);
```

$$Q := \begin{bmatrix} \frac{1}{2} & \frac{1}{2}\sqrt{2} & \frac{1}{6}\sqrt{6} & -\frac{1}{6}\sqrt{3} \\ \frac{-1}{2} & \frac{1}{2}\sqrt{2} & -\frac{1}{6}\sqrt{6} & \frac{1}{6}\sqrt{3} \\ \frac{-1}{2} & 0 & \frac{1}{3}\sqrt{6} & \frac{1}{6}\sqrt{3} \\ \frac{1}{2} & 0 & 0 & \frac{1}{2}\sqrt{3} \end{bmatrix}$$

```
> evalm(Q &* diag(sqrt(6),sqrt(2),sqrt(2),sqrt(2)));
```

$$\begin{bmatrix} \frac{1}{2}\sqrt{6} & 1 & \frac{1}{6}\sqrt{6}\sqrt{2} & -\frac{1}{6}\sqrt{3}\sqrt{2} \\ -\frac{1}{2}\sqrt{6} & 1 & -\frac{1}{6}\sqrt{6}\sqrt{2} & \frac{1}{6}\sqrt{3}\sqrt{2} \\ -\frac{1}{2}\sqrt{6} & 0 & \frac{1}{3}\sqrt{6}\sqrt{2} & \frac{1}{6}\sqrt{3}\sqrt{2} \\ \frac{1}{2}\sqrt{6} & 0 & 0 & \frac{1}{2}\sqrt{3}\sqrt{2} \end{bmatrix}$$

Figure 15.9: Orthogonal diagonalization of A

15.3   Exercises

1. Determine which of the following matrices are diagonalizable. For each that can be diagonalized, exhibit the diagonalization and show all your work.

$$A := \begin{bmatrix} 1 & -3 & -4 & 4 & -2 \\ 10 & 12 & 4 & 0 & 12 \\ 2 & 1 & 1 & 2 & 2 \\ 5 & 5 & 2 & 2 & 6 \\ -9 & -7 & 0 & -4 & -8 \end{bmatrix} \qquad B := \begin{bmatrix} 6 & 8 & -12 & 6 & -2 \\ 4 & 4 & -8 & 6 & -2 \\ 8 & 10 & -16 & 9 & -3 \\ 4 & 6 & -8 & 4 & -2 \\ 4 & 6 & -8 & 5 & -3 \end{bmatrix}$$

$$C := \begin{bmatrix} 16 & 6 & -1 & 7 & 19 \\ -9 & -1 & 1 & -5 & -13 \\ 5 & 2 & 2 & 3 & 7 \\ -7 & -3 & 1 & -1 & -10 \\ -4 & -2 & 0 & -2 & -3 \end{bmatrix}$$

2. Find a cube root of each of the following matrices, with real entries. *Check your final answers!*

$$A := \begin{bmatrix} 4 & 0 & -1 & 1 & 2 \\ -3 & 2 & 3 & 0 & -3 \\ 9 & 0 & -4 & 3 & 6 \\ -5 & 0 & 1 & -2 & -2 \\ 4 & 0 & -2 & 2 & 3 \end{bmatrix} \qquad B := \begin{bmatrix} -2 & -4 & 6 & -4 & 2 \\ -2 & -1 & 4 & -4 & 2 \\ -4 & -5 & 9 & -6 & 3 \\ -2 & -3 & 4 & -2 & 2 \\ -2 & -3 & 4 & -4 & 4 \end{bmatrix}$$

$$C := \begin{bmatrix} 66 & -51 & -12 & 8 & -4 \\ 58 & -43 & -12 & 8 & -4 \\ 96 & -72 & -19 & 12 & -6 \\ 58 & -42 & -12 & 7 & -4 \\ 67 & -51 & -12 & 8 & -5 \end{bmatrix}$$

3. For students who have studied calculus. Find the limit as n goes to infinity (if it exists) of the nth power of each of the following matrices.

$$A := \begin{bmatrix} \frac{1}{2} & -\frac{1}{2} & \frac{1}{2} & -\frac{1}{2} & \frac{1}{2} \\ -\frac{1}{2} & -\frac{1}{4} & \frac{5}{4} & -\frac{5}{4} & \frac{1}{2} \\ \frac{1}{6} & -\frac{1}{2} & \frac{5}{6} & -\frac{1}{2} & \frac{1}{2} \\ \frac{2}{3} & 0 & -\frac{2}{3} & 1 & 0 \\ 0 & -\frac{3}{4} & \frac{3}{4} & -\frac{3}{4} & 0 \end{bmatrix} \qquad B := \begin{bmatrix} 1 & 9 & -9 & 13 & -13 \\ \frac{1}{2} & -\frac{11}{2} & 6 & -\frac{19}{2} & \frac{19}{2} \\ \frac{1}{2} & 9 & -\frac{17}{2} & 13 & -13 \\ -3 & 0 & 3 & -1 & 0 \\ -3 & -9 & 12 & -15 & 14 \end{bmatrix}$$

4. Orthogonally diagonalize the following matrices.

$$A := \begin{bmatrix} 1 & 0 & 2 & 2 \\ 0 & 1 & 2 & -2 \\ 2 & 2 & 1 & 0 \\ 2 & -2 & 0 & 1 \end{bmatrix} \qquad B := \begin{bmatrix} 4 & 0 & 1 & 1 \\ 0 & 4 & 1 & -1 \\ 1 & 1 & 5 & 0 \\ 1 & -1 & 0 & 5 \end{bmatrix}$$

5. For the matrices of Exercise 4 above, find matrices E and F so that $EE^t = A$ and $FF^t = B$.

6. Orthogonally diagonalize the following matrices.

$$A := \begin{bmatrix} x & -1 & -1 & 1 \\ -1 & x & 1 & -1 \\ -1 & 1 & x & -1 \\ 1 & -1 & -1 & x \end{bmatrix} \qquad B := \begin{bmatrix} 1 & 1 & 1 & x \\ 1 & 1 & x & 1 \\ 1 & x & 1 & 1 \\ x & 1 & 1 & 1 \end{bmatrix}$$

7. A is a 5 by 5 matrix with eigenvalues 1 and 4. A basis for the eigenspace of A associated with 1 is $\{(2,4,3,1,5),(3,5,1,2,2),(1,1,2,4,1)\}$, and a basis for the eigenspace of A associated with 4 is $\{(3,1,3,3,3),(4,2,1,5,2)\}$. Find A.

15.4   Exploration and Discovery

1. In Chapter 3 we approximated transcendental functions evaluated at a matrix. We are now in a position to make exact calculations. Let $f(x) = \sum_{n=0}^{\infty} a_n x^n$ be a function given by a power series. If A is a matrix, we define $f(A)$ to be $\sum_{n=0}^{\infty} a_n A^n$, provided that the series converges. Observe that for diagonal matrices D the calculation of $f(D)$ is the same as the evaluation of f at each of the diagonal entries. Thus, for example,

$$\sin\left(\begin{bmatrix} 1 & 0 & 0 \\ 0 & 2 & 0 \\ 0 & 0 & 3 \end{bmatrix}\right) = \begin{bmatrix} \sin(1) & 0 & 0 \\ 0 & \sin(2) & 0 \\ 0 & 0 & \sin(3) \end{bmatrix}.$$

Furthermore, if $A = PDP^{-1}$, then we have the following:

$$\begin{aligned} f(A) &= \sum_{n=0}^{\infty} a_n A^n = \sum_{n=0}^{\infty} a_n (PDP^{-1})^n = \sum_{n=0}^{\infty} a_n PD^n P^{-1} \\ &= P(\sum_{n=0}^{\infty} a_n D^n)P^{-1} = P\, f(D)\, P^{-1} \end{aligned}$$

(a) If $A := \begin{bmatrix} 2 & 1 & -1 \\ 1 & 2 & -1 \\ -1 & 1 & 2 \end{bmatrix}$, $D := \begin{bmatrix} 1 & 0 & 0 \\ 0 & 2 & 0 \\ 0 & 0 & 3 \end{bmatrix}$, and $P := \begin{bmatrix} 1 & 1 & 1 \\ 0 & 1 & 1 \\ 1 & 1 & 0 \end{bmatrix}$, verify that $A = PDP^{-1}$.

(b) Calculate $\sin(A) = P \sin(D)P^{-1}$ and use approx to approximate it.

(c) Use Maclaurin polynomials of degrees 9 and 17 for $\sin(x)$ to approximate $\sin(A)$ as we did in Chapter 3. Compare your results with those in (b).

(d) Repeat all the above for $\cos(x)$ and e^x.

(e) For $A := \begin{bmatrix} 16 & 6 & 3 & 15 \\ 12 & 4 & 3 & 15 \\ 36 & 18 & 7 & 45 \\ -24 & -12 & -6 & -32 \end{bmatrix}$, calculate $\sin(A)$, $\cos(A)$, and e^A.

2. A matrix is said to be *nilpotent* if some power of A is the zero matrix. Let

$$A := \begin{bmatrix} 0 & p & q & r \\ 0 & 0 & s & t \\ 0 & 0 & 0 & u \\ 0 & 0 & 0 & 0 \end{bmatrix} \quad \text{and let} \quad B := \begin{bmatrix} 0 & p & q & r & s \\ 0 & 0 & t & u & v \\ 0 & 0 & 0 & w & x \\ 0 & 0 & 0 & 0 & y \\ 0 & 0 & 0 & 0 & 0 \end{bmatrix}.$$ Show that A and B are

nilpotent. What can you say about a matrix C all of whose entries on the main diagonal and below are zero? Which nilpotent matrices are diagonalizable?

15.5   Diagonalizing a Matrix

LABORATORY EXERCISE 15–1

Name _____   Due Date _____

Let $A :=
\begin{bmatrix}
1 & -2 & 4 & 2 \\
-2 & 1 & 4 & 2 \\
0 & -2 & 5 & 2 \\
-2 & -2 & 4 & 5
\end{bmatrix}$.

1. Find the characteristic polynomial of A.

2. Find the exact eigenvalues of A.

3. Find a basis for each eigenspace of A.

4. Find a matrix P such that $P^{-1}AP$ is diagonal.

5. Let D be the diagonal matrix in 4 above. Find AD. How is AD related to P? Explain.

Chapter 16
MATRICES AND LINEAR TRANSFORMATIONS
FROM $\mathbb{R}^M$ TO $\mathbb{R}^N$

LINEAR ALGEBRA CONCEPTS

- **Linear transformation**
- **Matrix of a linear transformation**
- **Kernel of a linear transformation**
- **Image of a linear transformation**
- **Inverse of a linear transformation**
- **Composition of linear transformations**

16.1 Introduction

Geometrically, linear transformations between vector spaces preserve straight lines. Algebraically, they preserve addition and scalar multiplication—the linear structure of the space. A linear transformation from $\mathbb{R}^m$ to $\mathbb{R}^n$ has a particularly simple description as a matrix product.

We will use the notation $[T]$ for the matrix of T with respect to the standard basis and call it the *standard matrix* of T. The matrix is defined as the n by m matrix whose ith column is $T(\mathbf{e}_i)$, where $\{\mathbf{e}_1, \mathbf{e}_2, \ldots, \mathbf{e}_m\}$ denotes the standard basis for $\mathbb{R}^m$.

The following theorem summarizes many properties of the standard matrix of a linear transformation and will be useful throughout this chapter.

Theorem: Let $T : \mathbb{R}^m \to \mathbb{R}^n$ be a linear transformation.

1. For each vector $\mathbf{v}$ in $\mathbb{R}^m$, $T(\mathbf{v}) = [T]\mathbf{v}$.

2. The kernel of T is the nullspace of $[T]$.

3. The image of T is the column space of $[T]$.

4. If $m = n$, then the standard matrix of T^k is $[T]^k$.

5. If $m = n$, then T has an inverse if and only if $[T]$ is invertible, in which case the standard matrix of T^{-1} is $[T]^{-1}$.

16.2   Solved Problems

<u>Solved Problem 1</u>: Let $T : \mathbb{R}^3 \to \mathbb{R}^4$ be defined by

$$T := (x, y, z) \to (x + 4y - 2z, 2x + y - z, 3x + 5y - 3z, -x + 3y - z)$$

and let $S : \mathbb{R}^2 \to \mathbb{R}^1$ be defined by

$$S := (x, y) \to \begin{vmatrix} x & y \\ x + y & x - y \end{vmatrix}$$

Determine if T and S are linear transformations.

<u>Solution</u>: After loading the `linalg` package, define T by

```
T := (x,y,z) -> [x+4*y-2*z, 2*x+y-z, 3*x+5*y-3*z, -x+3*y-z]
```

(<u>Note</u>: We are abusing the notation and using the list `[a,b,c]` in place of `vector([a,b,c])`. Conveniently, Maple will accept it in this problem.) T is a linear transformation provided that it satisfies the following two conditions for all (x, y, z) and (u, v, w) in $\mathbb{R}^3$ and all $\alpha \in \mathbb{R}$.

$$\begin{aligned} T\left((x, y, z) + (u, v, w)\right) &= T(x, y, z) + T(u, v, w) \\ T\left(\alpha\,(x, y, z)\right) &= \alpha T(x, y, z) \end{aligned}$$

or equivalently,

$$\begin{aligned} T\left((x, y, z) + (u, v, w)\right) - T(x, y, z) - T(u, v, w) &= 0 \\ T\left(\alpha\,(x, y, z)\right) - \alpha T(x, y, z) &= 0 \end{aligned}$$

Thus, using the first conditions, we enter

```
equal( T(x+u,y+v,z+w), T(x,y,z) + T(u,v,w))
```

to test if the two vectors are the same. Maple responds with *true*. The `equal` function from `linalg` is for testing matrices and vectors. Using the equivalent form of the conditions, evaluate `T(c*x,c*y,c*z) - c*T(x,y,z)` as a matrix/vector expression. From lines 3 and 4 of Figure 16.1, we see that $T\left(c\,(x, y, z)\right) - cT(x, y, z)$ simplifies to the zero vector. (<u>Note</u>: To simplify the elements of the vector, we mapped `expand` in line 4.) We conclude that T is a linear transformation.

The second part of the exercise is easy to do by hand and certainly does not require Maple, but we present this nonlinear case for comparison.

Define S with `S := (x,y) -> det(matrix([[x,y], [x+y,x-y]]))`. Proceeding as above, enter and simplify `S(x+u,y+v) - S(x,y) - S(u,v)`. The result displayed in line 6 of Figure 16.1 does not appear to be zero, but to verify it you should select specific values of x, y, u, v that do not yield 0. Choosing $x = y = u = v = 1$ gives $S(1+1, 1+1) - S(1,1) - S(1,1) = -2 \neq 0$, so S is not a linear transformation.

```
> T := (x,y,z) -> [x+4*y-2*z,2*x+y-z,3*x+5*y-3*z,-x+3*y-z]:
> equal(T(x+u,y+v,z+w), T(x,y,z) + T(u,v,w));
```

$$true$$

```
> evalm(T(c*x,c*y,c*z) - c*T(x,y,z));
```

$$[cx + 4cy - 2cz - c(x + 4y - 2z) \quad 2cx + cy - cz - c(2x + y - z)$$
$$3cx + 5cy - 3cz - c(3x + 5y - 3z) \quad -cx + 3cy - cz - c(-x + 3y - z)]$$

```
> map(expand, ");
```

$$[0 \quad 0 \quad 0 \quad 0]$$

```
> S := (x,y) -> det(matrix([[x,y], [x+y,x-y]])):
> simplify(S(x+a,y+b) - S(x,y) - S(a,b));
```

$$2xa - 2xb - 2ay - 2yb$$

Figure 16.1: | Testing functions to determine if they are linear transformations |

Solved Problem 2: Let $T : \mathbb{R}^2 \to \mathbb{R}^2$ be defined by

$$T := (x, y) \to (x + 4y, 2x + y)$$

and let $S : \mathbb{R}^2 \to \mathbb{R}^2$ be defined by

$$S := (x, y) \to (xy, 2x + y)$$

Examine the images under T and S of two lines as a test for linearity: a linear transformation must map lines to lines.

Solution: Define `T := (x,y) -> (x+4y,2x+y)` as seen in Figure 16.2. (We used parentheses instead of brackets to make the syntax easier for parametric plotting.)

Let's consider the line $y = x$. To graph the image of $y = x$ under T requires parametric plotting. Maple's basic syntax for parametric plotting is

$$\texttt{plot(} [x(t), y(t), t = t_{initial}..t_{final}] \texttt{)}$$

Note the brackets around both the function and the parameter interval. (This is why we defined T with parentheses instead of brackets or as a vector.)

To plot the line, replace x and y by t to yield `T(t,t)`, and use the range `t=-3..3`. The graph is a straight line. This is repeated for $x = 2y$ in line 3 of Figure 16.2; we see a second line. (To save space we have merged the two graphs with `display`.) Thus, we suspect T is linear because it transforms lines to lines. To *prove* it's linear we must verify the property $T(a\mathbf{x} + \mathbf{y}) = aT(\mathbf{x}) + T(\mathbf{y})$.

```
> T := (x,y) -> (x+4y,2x+y):
> plot( [T(t,t), t=-3..3] );
> plot( [T(2*t,t), t=-3..3] );
```

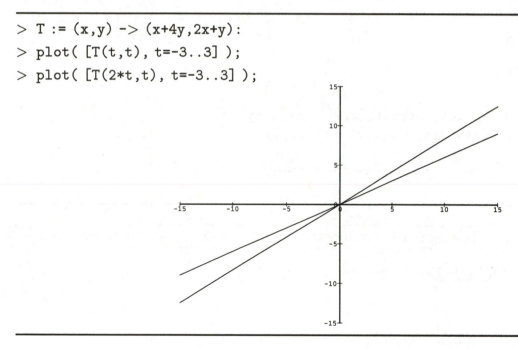

Figure 16.2: | Image of lines under a linear transformation |

Enter S with `S := (x,y) -> (x*y,2*x+y)` as seen in line 1 of Figure 16.3. Let's consider the line $y = x$. As above, replace x and y by t and plot for `t=-3..3`. The graph is a curve, actually, a parabola. Repeat this for $x = 2y$ to see a second parabola. This shows that S transforms straight lines into curves that are not lines. Thus, S is nonlinear.

```
> S := (x,y) -> (x*y,2*x+y):
> plot( [S(t,t), t=-3..3] );
> plot( [S(2*t,t), t=-3..3] );
```

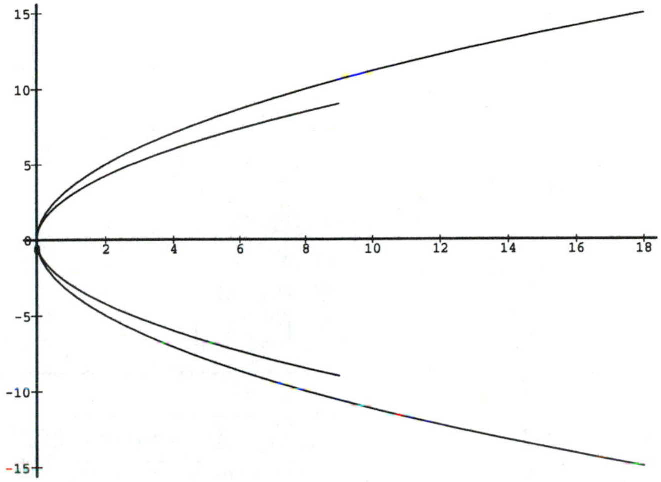

Figure 16.3: Image of lines under a non-linear transformation

Solved Problem 3: Let T be the linear transformation from Solved Problem 1. Find the standard matrix of T and find bases for the image and kernel of T.

Solution: To get the standard matrix of T, we can enter T := (x,y,z) -> [x+4*y-2*z, 2*x+y-z, 3*x+5*y-3*z, -x+3*y-z] and then define $[T]$ by MT := augment(T(1,0,0), T(0,1,0), T(0,0,1)) as seen in Figure 16.4. The standard matrix $[T]$ of T appears in line 2 of Figure 16.4.

The kernel of T is the null space of $[T]$, and the image of T is the column space of $[T]$. In Chapter 10 we found bases for both the null space and column space of a matrix, so we simply repeat the process here.

Enter rref(MT). From the reduced form of the matrix displayed in line 3 of Figure 16.4, we read a basis for the kernel of T: $\{(\frac{2}{7}, \frac{3}{7}, 1)\}$. Line 3 of Figure 16.4 shows the first two columns of the reduced form of $[T]$ contain leading ones. Thus, the first two columns of $[T]$ form a basis for the image of T: $\{(1, 2, 3, -1), (4, 1, 5, 3)\}$.

```
> T := (x,y,z) -> [x+4*y-2*z, 2*x+y-z, 3*x+5*y-3*z, -x+3*y-z]:
> MT := augment(T(1,0,0), T(0,1,0), T(0,0,1));
```

$$MT := \begin{bmatrix} 1 & 4 & -2 \\ 2 & 1 & -1 \\ 3 & 5 & -3 \\ -1 & 3 & -1 \end{bmatrix}$$

```
> rref(MT);
```

$$\begin{bmatrix} 1 & 0 & \frac{-2}{7} \\ 0 & 1 & \frac{-3}{7} \\ 0 & 0 & 0 \\ 0 & 0 & 0 \end{bmatrix}$$

Figure 16.4: $\boxed{\text{Kernel and image of a linear transformation}}$

Solved Problem 4: Let $T : \mathbb{R}^3 \to \mathbb{R}^3$ be defined by $T := (x, y, z) \to (x - 2y + 3z, 2x + y - z, x + 3y + 2z)$.

1. Find the standard matrix of T.

2. Find $T^3(x, y, z)$.

3. Show that T has an inverse and find $T^{-1}(x, y, z)$.

Solution: We could just jot down the standard matrix on paper then enter it, or we can define T by `T := (x,y,z) -> [x-2y+3z, 2x+y-z, x+3y+2z]` and then enter `MT := augment(T(1,0,0), T(0,1,0), T(0,0,1))`.

Since $T^3(x) = [T]^3 x$, calculate `MT^3 &* [x,y,z]`. We conclude from line 3 in Figure 16.5 that $T^3(x, y, z) = [21x + 38y + 17z, -6x - 3y + 21z, 27x + y + 28z]$.

The linear transformation T has an inverse if and only if the matrix $[T]$ does, in which case $T^{-1}(x) = [T]^{-1} x$. Therefore, we evaluate `MT^(-1) &* [x,y,z]`. (Note the space after the 1.) We conclude from line 6 in Figure 16.5 that the matrix $[T]$ has an inverse and that $T^{-1}(x, y, z) = [x/6 + 13y/30 - z/30, \ -x/6 - y/30 + 7z/30, \ x/6 - y/6 + z/6]$. (If the matrix $[T]$ did not have an inverse, line 4 would yield an error.)

```
> T := (x,y,z) -> [x-2y+3z, 2x+y-z, x+3y+2z]:
> MT := augment(T(1,0,0), T(0,1,0), T(0,0,1));
```

$$MT := \begin{bmatrix} 1 & -2 & 3 \\ 2 & 1 & -1 \\ 1 & 3 & 2 \end{bmatrix}$$

```
> evalm(MT^3 &* vector([x,y,z]));
```

$$[21x + 38y + 17z \quad -6x - 3y + 21z \quad 27x + y + 28z]$$

```
> evalm( MT^(-1) &* vector([x,y,z]));
```

$$\left[\frac{1}{6}x + \frac{13}{30}y - \frac{1}{30}z \quad -\frac{1}{6}x - \frac{1}{30}y + \frac{7}{30}z \quad \frac{1}{6}x - \frac{1}{6}y + \frac{1}{6}z\right]$$

Figure 16.5: Inverse and power of a linear transformation

16.3   Exercises

1. Determine if the following are linear transformations.

   (a) $T : \mathbb{R}^3 \to \mathbb{R}^3$, $T := (x, y, z) \to (3x - 5y + 4z, 8x - 7y - 4z, 2x + 5y - \pi z)$

   (b) $T : \mathbb{R}^3 \to \mathbb{R}^3$, $T := (x, y, z) \to ((x - y)^2, (x - z)^2, (y - z)^2)$

   (c) $T : \mathbb{M}_{2,2} \to \mathbb{M}_{2,2}$, $T := A \to A^2 - 3A$

2. Repeat Solved Problem 2 for $T := (x, y) \to (x^2, y^2)$ and for $S := (x, y) \to \left(x \sin^2(y), x \cos^2(y)\right)$. Discuss your observations and conclusions. Are T and S linear?

3. For each of the following linear transformations T, find the standard matrix of T and find bases for both the image and kernel of T.

   (a) $T : \mathbb{R}^4 \to \mathbb{R}^3$, $T := (x, y, z, w) \to (2x - 3y + 4z - w, 5x - y + 2z + 4w, 3x + 2y - 2z + 5w)$

   (b) $T : \mathbb{R}^4 \to \mathbb{R}^3$, $T := \mathbf{v} \to A\mathbf{v}$, where $A := \begin{bmatrix} 2 & 3 & 1 & 1 \\ 3 & 2 & 3 & 3 \\ 5 & 5 & 4 & 4 \end{bmatrix}$.

4. For each of the following linear transformations T, find $T^5(x, y, z, w)$. Determine if T^{-1} exists. If it does, find $T^{-1}(x, y, z, w)$.

   (a) $T : \mathbb{R}^4 \to \mathbb{R}^4$, $T := (x, y, z, w) \to (x - y - z, x + 2y + 3w, x + 4y + z + w, 3x - 2y + z - 2w)$.

   (b) $T : \mathbb{R}^4 \to \mathbb{R}^4$, $T := \mathbf{v} \to \begin{bmatrix} 1 & 1 & 2 & 4 \\ 2 & 1 & 3 & 2 \\ 4 & 2 & 1 & 3 \\ 7 & 2 & 3 & 1 \end{bmatrix} \mathbf{v}$.

16.4   Matrices and Linear Transformations

LABORATORY EXERCISE 16–1

Name _____   **Due Date** _____

Let $T : \mathbb{R}^4 \rightarrow \mathbb{R}^4$, $T := (x, y, z, w) \rightarrow (2x + z - w, -y + z + w, x + y - 2w, x - 2w)$.

1. Find the standard matrix of T.

2. Find a basis for the image of T.

3. Find a basis for the kernel of T.

4. Determine if T^{-1} exists. If it does, find $T^{-1}(x, y, z, w)$.

Chapter 17
MATRICES OF GENERAL LINEAR TRANSFORMATIONS; SIMILARITY

LINEAR ALGEBRA CONCEPTS

- **Matrix of a linear transformation**
- **Similar matrices**

17.1   Introduction

A linear transformation between finite-dimensional vector spaces can be represented by matrix multiplication once ordered bases have been selected for the two vector spaces. A given linear transformation can be represented by different matrices depending on the chosen bases.

17.2   Solved Problems

Solved Problem 1: Let $T : \mathbb{R}^3 \to \mathbb{R}^4$ be the linear transformation defined by $T(x, y, z) = (2x + 3y - 5z, 3x + y + 4z, 2x - 3y + z, x - y + 2z)$, let $B = \{(1, 1, 2), (2, 1, 1), (3, 0, 1)\}$ and let $B' = \{(2, 1, 1, 1), (1, 3, 2, 1), (2, 1, 1, 2), (1, 3, 3, 2)\}$. Find the matrix of T with respect to the bases B and B'.

<u>Solution</u>: Recall that the columns of A are the B' coordinates of the images of elements of B. Thus, if $[T]$ is the standard matrix of T, M is the matrix whose columns are the vectors in B, and N is the matrix whose columns are the vectors in B', then $A = N^{-1}[T]M$. The easiest way to enter the standard matrix is to define T and then apply T to the standard basis vectors e_1, e_2, and e_3 as shown in the following:

```
T := (x,y,z) -> [2*x+3*y-5*z, 3*x+y+4*z, 2*x-3*y+z, x-y+2*z]
Mt:= augment(T(1,0,0), T(0,1,0), T(0,0,1))
```

(Notice that we have abused the notation by not using Maple's **vector**; in this case, Maple is forgiving. See Figure 17.1.) The standard matrix of T appears in line 1.

Next enter the two matrices M and N given below.

$$M := \begin{bmatrix} 1 & 2 & 3 \\ 1 & 1 & 0 \\ 2 & 1 & 1 \end{bmatrix} \qquad N := \begin{bmatrix} 2 & 1 & 2 & 1 \\ 1 & 3 & 1 & 3 \\ 1 & 2 & 1 & 3 \\ 1 & 1 & 2 & 2 \end{bmatrix}$$

These two matrices were entered as the transpose of the vectors of the basis; this was easier typing. Also, to save space, we printed them together instead of on separate lines. These two matrices appear in line 2 of Figure 17.1. Finally, evaluate $A := N^{-1} Mt\, M$. The matrix A of T with respect to B and B' is in line 3.

```
> T := (x,y,z) -> [2x+3y-5z, 3x+y+4z, 2x-3y+z, x-y+2z]:
  Mt := augment(T(1,0,0), T(0,1,0), T(0,0,1));
```

$$Mt := \begin{bmatrix} 2 & 3 & -5 \\ 3 & 1 & 4 \\ 2 & -3 & 1 \\ 1 & -1 & 2 \end{bmatrix}$$

```
> M := transpose(matrix(3,3, [1,1,2, 2,1,1, 3,0,1])):
  N := transpose(matrix(4,4,[2,1,1,1, 1,3,2,1, 2,1,1,2, 1,3,3,2])):
  print(M,N);
```

$$\begin{bmatrix} 1 & 2 & 3 \\ 1 & 1 & 0 \\ 2 & 1 & 1 \end{bmatrix}, \begin{bmatrix} 2 & 1 & 2 & 1 \\ 1 & 3 & 1 & 3 \\ 1 & 2 & 1 & 3 \\ 1 & 1 & 2 & 2 \end{bmatrix}$$

```
> A := evalm(N^(-1) &* Mt &* M);
```

$$\begin{bmatrix} \frac{-71}{5} & -6 & -5 \\ 11 & 9 & 6 \\ \frac{44}{5} & 5 & 3 \\ \frac{-26}{5} & -5 & -1 \end{bmatrix}$$

Figure 17.1: A nonstandard matrix of a linear transformation

Solved Problem 2: Let $T : \mathbb{P}_3 \to \mathbb{M}_{2,2}$ be the linear transformation defined by

$$T\,(p) := \begin{bmatrix} p(1) & p(2) \\ p(1)+p(2) & p(1)-p(2) \end{bmatrix}$$

1. Find the standard matrix of T.

2. Find a basis for the kernel of T.

3. Find a basis for the image of T.

Solution: Recall that the standard basis for $\mathbb{P}_3$ is $B = \{x^3, x^2, x, 1\}$ and that the standard basis for $\mathbb{M}_{2,2}$ is

$$B' = \left\{ \begin{bmatrix} 1 & 0 \\ 0 & 0 \end{bmatrix}, \begin{bmatrix} 0 & 1 \\ 0 & 0 \end{bmatrix}, \begin{bmatrix} 0 & 0 \\ 1 & 0 \end{bmatrix}, \begin{bmatrix} 0 & 0 \\ 0 & 1 \end{bmatrix} \right\}$$

To find the standard matrix A of T, we need the B' coordinates of:

$$T\left(x^3\right) = \begin{bmatrix} 1 & 8 \\ 9 & -7 \end{bmatrix}, \ T\left(x^2\right) = \begin{bmatrix} 1 & 4 \\ 5 & -3 \end{bmatrix}, \ T(x) = \begin{bmatrix} 1 & 2 \\ 3 & -1 \end{bmatrix}, \ T(1) = \begin{bmatrix} 1 & 1 \\ 2 & 0 \end{bmatrix}$$

The coordinates are found easily by inspection. The B' coordinates of these vectors go into the columns of the standard matrix,

$$A = \begin{bmatrix} 1 & 1 & 1 & 1 \\ 8 & 4 & 2 & 1 \\ 9 & 5 & 3 & 2 \\ -7 & -3 & -1 & 0 \end{bmatrix}.$$

Enter the matrix A and then row-reduce it using the **rref** function. From line 2 of Figure 17.2 we read a basis for the nullspace of A as $\left\{(\frac{1}{2}, -\frac{3}{2}, 1, 0), (\frac{3}{4}, -\frac{7}{4}, 0, 1)\right\}$. These are the B coordinates of a basis for the kernel of T. Thus, a basis for the kernel of T is $\left\{\frac{1}{2}x^3 - \frac{3}{2}x^2 + x, \ \frac{3}{4}x^3 - \frac{7}{4}x^2 + 1\right\}$.

Similarly, a basis for the column space of A is $\{(1, 8, 9, -7), (1, 4, 5, -3)\}$. These are the B' coordinates of a basis for the image of T. Thus, a basis for the image of T is

$$\left\{ \begin{bmatrix} 1 & 8 \\ 9 & -7 \end{bmatrix}, \begin{bmatrix} 1 & 4 \\ 5 & -3 \end{bmatrix} \right\}$$

```
> A := matrix(4,4,[1,1,1,1, 8,4,2,1, 9,5,3,2, -7,-3,-1,0]);
```

$$A := \begin{bmatrix} 1 & 1 & 1 & 1 \\ 8 & 4 & 2 & 1 \\ 9 & 5 & 3 & 2 \\ -7 & -3 & -1 & 0 \end{bmatrix}$$

```
> rref(A);
```

$$\begin{bmatrix} 1 & 0 & \frac{-1}{2} & \frac{-3}{4} \\ 0 & 1 & \frac{3}{2} & \frac{7}{4} \\ 0 & 0 & 0 & 0 \\ 0 & 0 & 0 & 0 \end{bmatrix}$$

Figure 17.2: | Kernel and image of a linear transformation |

Solved Problem 3: Let $A = \begin{bmatrix} 4 & 1 & -1 \\ -2 & 5 & 2 \\ 4 & -4 & 1 \end{bmatrix}$, $B = \begin{bmatrix} 3 & 3 & 1 \\ -1 & -1 & -1 \\ 1 & 10 & 6 \end{bmatrix}$, and $C = \begin{bmatrix} 3 & 3 & 0 \\ -1 & 3 & 1 \\ 1 & 1 & 2 \end{bmatrix}$.

1. Determine if A and B are similar.

2. Determine if B and C are similar.

Solution: Enter the matrices A and B and name them as seen in Figure 17.3. Also enter P as a matrix with undefined entries as seen in line 2 of Figure 17.3.

We seek an invertible matrix P such that $PAP^{-1} = B$. This matrix equation is very difficult for Maple to deal with; in this small 3×3 problem, to expand PAP^{-1} requires 180 additions, 9 divisions, and 504 multiplications! So we *condition* the equation by multipling both sides on the right by P to get $PA = BP$. Notice that, unlike the first equation, this one, while much easier to solve, may have noninvertible solutions—the zero matrix, for example.

Define the equation eq := P &* A = B &* P as seen in line 3 of Figure 17.3. (Recall that we need the function matsolve from Appendix B for matrix equations.) If we solve the resulting system of equations, we see from line 4 of Figure 17.3 that the only solution of the equation $PA = BP$ is the zero matrix. Since the zero matrix is not invertible, we conclude that the equation $PAP^{-1} = B$ has no solution and hence the matrices A and B are not similar.

```
>  A := matrix(3,3,[4,1,-1, -2,5,2, 4,-4,1]):
   B := matrix(3,3,[3,3,1, -1,-1,-1, 1,10,6]):
> P := matrix(3,3):  evalm(P);
```

$$P := \begin{bmatrix} P_{1,1} & P_{1,2} & P_{1,3} \\ P_{2,1} & P_{2,2} & P_{2,3} \\ P_{3,1} & P_{3,2} & P_{3,3} \end{bmatrix}$$

```
> eq := (P &* A)=(B &* P);
```

$$eq := P \text{ \&* } A = B \text{ \&* } P$$

```
> matsolve(eq);
```

$$\{P_{1,1} = 0, P_{1,2} = 0, P_{1,3} = 0, P_{2,1} = 0, P_{2,1} = 0, P_{2,2} = 0, P_{2,3} = 0,$$
$$P_{3,1} = 0, P_{3,2} = 0, P_{3,3} = 0\}$$

Figure 17.3: Matrices that are not similar

Remark: There is an easier way to solve this problem. Recall that similar matrices must have the same determinant. Verify that the determinants of A and B are different, and hence the matrices are not similar. However, if the determinants were the same, we would have to go through this procedure.

To solve the second part of the problem, enter the matrix C and name it as in Figure 17.4. As we did before, define eq := P &* A = B &* P and solve. From line 3 of Figure 17.4, we see that the equation $PB = CP$ has infinitely many solutions. We need only choose one that yields an invertible matrix P. In line 4, we use assign('') and evalm(P) to insert the solution into P. (The equations in the solution set are treated as definitions by assign.) Next, use subs to set $P_{1,2} = 1$, $P_{1,3} = 2$, and $P_{2,3} = 3$.

(There are many other choices that will lead to a correct answer.) We see the resulting matrix in line 5 of Figure 17.4. This choice of parameters leads to the solution

$$P = \begin{bmatrix} -2 & 1 & 2 \\ 1 & 10 & 3 \\ -9 & -6 & 2 \end{bmatrix}$$

Notice that in lines 13 and 14 of Figure 17.4 we have checked to see that the determinant of P is not zero, thus ensuring that P has an inverse. We conclude that C is similar to B.

```
> C := matrix(3,3,[3,1,0, -1,3,1, 1,1,2]):
> eq := P &* A = B &* P:
> matsolve(eq);
```

$$\{P_{2,1} = -P_{1,2} + P_{1,3}, P_{3,2} = 2P_{1,2} - P_{1,3} - 2P_{2,3}, P_{3,1} = 2P_{1,2} - 4P_{1,3} - P_{2,3},$$
$$P_{3,3} = P_{1,3}, P_{2,3} = P_{2,3}, P_{1,3} = P_{1,3}, P_{1,2} = P_{1,2}, P_{1,1} = P_{1,2} - 3P_{1,3} + P_{2,3},$$
$$P_{2,2} = -P_{1,2} + P_{1,3} + 3P_{2,3}\}$$

```
> assign("):  evalm(P);
```

$$\begin{bmatrix} P_{1,2} - 3P_{1,3} + P_{2,3} & P_{1,2} & P_{1,3} \\ -P_{1,2} + P_{1,3} & -P_{1,2} + P_{1,3} + 3P_{2,3} & P_{2,3} \\ 2P_{1,2} - 4P_{1,3} - P_{2,3} & 2P_{1,2} - P_{1,3} - 2P_{2,3} & P_{1,3} \end{bmatrix}$$

```
> P := subs(P[1,2]=1, P[1,3]=2, P[2,3]=3, ");
```

$$P := \begin{bmatrix} -2 & 1 & 2 \\ 1 & 10 & 3 \\ -9 & -6 & 2 \end{bmatrix}$$

```
> det(P);
```

$$63$$

Figure 17.4: | Matrices that are similar |

Solved Problem 4: Let $T(x, y, z, w) = (5x + y + z + 3w, \ 4x + 3y + 2z + 4w, \ 2x + 4z + 2w, \ -4x - y - 2z - 2w)$. Find a basis B for $\mathbb{R}^4$ so that the matrix of T with respect to the basis B is a diagonal matrix.

Solution: The matrix of T with respect to the basis B is diagonal if and only if B consists of eigenvectors of T. The procedure for obtaining eigenvectors from the standard matrix of T should be familiar. The standard matrix of T is

$$\begin{bmatrix} 5 & 1 & 1 & 3 \\ 4 & 3 & 2 & 4 \\ 2 & 0 & 4 & 2 \\ -4 & -1 & -2 & -2 \end{bmatrix}$$

Enter this matrix and name it A. To find the eigenvalues, use `charpoly(A, lambda)` and then `solve`. The eigenvalues of A are 2 and 3 as seen from line 2 of Figure 17.5. Enter `alias(Id=&*())` to define the identity matrix as Id.

To get the eigenvectors belonging to 2, enter `rref(2*Id - A)`. From the result in line 3, we read a basis for the eigenspace corresponding to 2 as $\{(-1, 2, 1, 0), (-1, 0, 0, 1)\}$. Similarly, for the eigenvalue 3 we read from line 4 the basis $\{(-\frac{1}{2}, 0, 1, 0), (-1, -1, 0, 1)\}$. Therefore, $B = \{(-1, 2, 1, 0), (-1, 0, 0, 1), (-\frac{1}{2}, 0, 1, 0), (-1, -1, 0, 1)\}$. Without further work, you should be able to give the matrix of T relative to B.

```
> p := charpoly(A,lambda);
```
$$p := \lambda^4 - 10\lambda^3 + 37\lambda^2 - 60\lambda + 36$$
```
> solve(p=0);
```
$$2, 2, 3, 3$$
```
> rref(2*Id - A);
```
$$\begin{bmatrix} 1 & 0 & 1 & 1 \\ 0 & 1 & -2 & 0 \\ 0 & 0 & 0 & 0 \\ 0 & 0 & 0 & 0 \end{bmatrix}$$
```
> rref(3*Id - A);
```
$$\begin{bmatrix} 1 & 0 & \frac{1}{2} & 1 \\ 0 & 1 & 0 & 1 \\ 0 & 0 & 0 & 0 \\ 0 & 0 & 0 & 0 \end{bmatrix}$$

Figure 17.5: | Diagonalizing a linear transformation |

17.3   Exercises

1. Let $T : \mathbb{R}^4 \to \mathbb{R}^3$ be the linear transformation defined by

$$T(x, y, z, w) = (x + y - z + w, \ x + 2y + 3z, \ 5x + 4y - 2w)$$

Let the bases B and B' be given by

$$B = \{(2, 1, 3, 3), \ (3, 2, 1, 4), \ (2, 2, 1, 5), \ (3, 3, 6, 4)\}$$
$$B' = \{(3, 1, 1), \ (2, 5, 5), \ (1, 2, 3)\}$$

Find the matrix of T with respect to B and B'.

2. Let B and B' be the bases from Exercise 1 above, and let $S : \mathbb{R}^4 \to \mathbb{R}^3$ be the linear transformation whose matrix with respect to B and B' is $\begin{bmatrix} 2 & 1 & 1 & 4 \\ 3 & 2 & 1 & 1 \\ 5 & 3 & 2 & 5 \end{bmatrix}$.
   Find $S(x, y, z, w)$.

3. Find bases for the kernel and image of the linear transformations T from Exercise 1 and S from Exercise 2.

4. **For students who have studied calculus.** Let $D : \mathbb{P}_3 \to \mathbb{P}_3$ be the linear transformation defined by $D(p) = \frac{dp}{dx}$ (the derivative of $p(x)$), and let $B = \{1, \ 1 + x, \ x + x^2, \ x^2 + x^3\}$. Find the matrix of D with respect to B.

5. Let $T : \mathbb{R}^3 \to M_{2,2}$ be the linear transformation defined by

$$T := (x, y, z) \to \begin{bmatrix} x + y & x - z \\ y + z & x + y + z \end{bmatrix}$$

Let

$$B = \{(1, 3, 3), (2, 1, 4), (3, 2, 5)\}$$
$$B' = \left\{ \begin{bmatrix} 2 & 1 \\ 3 & 1 \end{bmatrix}, \begin{bmatrix} 1 & 1 \\ 2 & 3 \end{bmatrix}, \begin{bmatrix} 3 & 2 \\ 1 & 3 \end{bmatrix}, \begin{bmatrix} 1 & 7 \\ 3 & 2 \end{bmatrix} \right\}$$

Find the matrix of T with respect to B and B'.

6. Let $T : \mathbb{P}_3 \to \mathbb{R}^3$ be defined by $T(p) = (p(1) + p(2), p(1) - p(2), p(1))$. Find bases for the image and kernel of T.

7. Let $T : \mathbb{R}^4 \to \mathbb{R}^4$ be the linear transformation defined by $T(x, y, z, w) = (11x + 4y + 4z + 12w, \ 18x + 4y + 8z + 18w, \ 12x + 2y + 7z + 12w, \ -18x - 5y - 8z - 19w)$. Find a basis B such that the matrix of T with respect to B is diagonal.

8. Let $T : \mathbb{P}_3 \to \mathbb{P}_3$ be defined by $T(ax^3 + bx^2 + cx + d) = (8a + 2b + 2c + 6d)x^3 + (2z + 3b + 2d + 2c)x^2 + (-2a - 2b + 5c - 2d)x + (-4a - b - 3c - 2d)$. Find a basis B for $\mathbb{P}_3$ such that the matrix of T with respect to B is diagonal.

9. Determine if the matrices $\begin{bmatrix} 1 & 2 & 1 & 1 & 3 \\ 2 & 1 & 4 & 2 & 3 \\ 1 & 2 & 3 & 1 & 2 \\ 3 & 1 & 1 & 2 & 3 \\ 2 & 2 & 2 & 3 & 1 \end{bmatrix}$ and $\begin{bmatrix} 0 & 6 & -1 & 8 & 0 \\ 1 & 3 & 0 & 6 & 2 \\ 0 & 8 & -1 & 11 & -1 \\ -1 & 5 & 0 & 5 & -1 \\ 2 & 5 & -2 & 6 & 1 \end{bmatrix}$ are similar.

10. Determine if the matrices $\begin{bmatrix} 1 & 2 & 3 & 2 \\ 1 & 2 & 4 & 2 \\ 2 & 2 & 1 & 1 \\ 2 & 1 & 2 & 3 \end{bmatrix}$ and $\begin{bmatrix} -215 & -90 & -60 & -252 \\ 216 & 91 & 60 & 252 \\ -432 & -180 & -19 & -504 \\ 215 & 90 & 60 & 253 \end{bmatrix}$ are similar.

17.4   Exploration and Discovery

1. Let $A = \begin{bmatrix} 11 & 4 & 4 & 12 \\ 18 & 4 & 8 & 18 \\ 12 & 2 & 7 & 12 \\ -18 & -5 & -8 & -19 \end{bmatrix}$, let $P = \begin{bmatrix} 1 & 1 & 1 & 0 \\ 0 & 1 & 1 & 1 \\ 1 & 0 & 1 & 1 \\ 1 & 1 & 0 & 1 \end{bmatrix}$, and let $B = PAP^{-1}$.

Then by definition, A is similar to B. Using A and B, test the following statements. Make up new matrices A and B of your own and test the statements a second time. Give counterexamples to those you determine to be false. Try to prove those that are true.

(a) Similar matrices have the same determinant.

(b) Similar matrices have the same reduced echelon form.

(c) Similar matrices have the same characteristic polynomial.

(d) Similar matrices have the same eigenvalues.

(e) Similar matrices have the same eigenspaces.

17.5   Matrices and Linear Transformations, I

LABORATORY EXERCISE 17–1

Name _____   **Due Date** _____

Let $T : \mathbb{P}_2 \to \mathbb{P}_2$ be a linear transformation such that

$$
\begin{aligned}
T(x^2 - x) &= 50x^2 - 21x + 65, \\
T(x^2) &= 5x^2 - 2x + 6, \\
T(x + 1) &= -26x^2 + 11x - 34
\end{aligned}
$$

1. Find the matrix, $[T]$, of T with respect to the standard basis of $\mathbb{P}_2$.

2. Find the matrix of T with respect to the basis $\{x^2 - x, x^2, x + 1\}$.

3. Determine if $[T]^{-1}$ exists.

4. Determine if T^{-1} exists. If it does, find $T^{-1}(ax^2 + bx + c)$.

17.6   Matrices and Linear Transformations, II

LABORATORY EXERCISE 17–2

Name _____   **Due Date** _____

Let $T : \mathbb{P}_3 \to \mathbb{P}_3$ be a linear transformation such that

$$
\begin{aligned}
T(x^3 - x) &= x^2 + 5, \\
T(x^3 + 1) &= x^2 - 1, \\
T(x^2) &= 3, \\
T(x^2 - 1) &= x^2 - x
\end{aligned}
$$

1. Find the matrix of T with respect to the standard basis of $\mathbb{P}_3$.

2. Find the matrix of T with respect to the basis $\{x^3 - x,\ x^3 + 1,\ x^2,\ x^2 - 1\}$.

3. Find a basis for the kernel of T.

4. Find a basis for the image of T.

Chapter 18
APPLICATIONS AND NUMERICAL METHODS

LINEAR ALGEBRA CONCEPTS

- **Systems of differential equations**
- **Gauss-Seidel method**
- **Generalized inverse and curve fitting**
- **Rotation of axes**
- LU **and** QR **factorizations**

18.1   Introduction

We began our study of linear algebra by solving systems of linear equations. In this chapter we look at how the ideas we have developed are used in solving systems of linear *differential* equations.

In some applications it may be necessary to solve huge systems of equations that are too large even for Maple to handle. (A computer's memory is finite.) In addition, the number of calculations required to row-reduce the augmented matrix for a very large system may require an impractical amount of time on even the fastest supercomputers and can lead to significant round-off error. For these reasons, other methods must be used to approximate solutions of large systems. We illustrate one such method known as *Gauss-Seidel iteration.*

Sometimes there may not be a solution to a system of equations, but there may be something very close to it. For example, there may not be a straight line passing through three given points, but there may be a line that comes close to all three points. We shall also explore this idea.

Certain operations on a matrix can result in a factorization of the matrix. It can also be advantageous to factor a matrix and perform calculations on the factors so as to gain large reductions in arithmetic complexity. We look at the two most common — the LU and QR factorizations.

18.2 Solved Problems

<u>Solved Problem 1</u>: | **For students who have studied calculus.** | Solve the following system of linear differential equations.

$$\frac{dx_1}{dt} = -8x_1 + 4x_2 + 9x_3$$

$$\frac{dx_2}{dt} = 33x_1 - 9x_2 - 27x_3$$

$$\frac{dx_3}{dt} = -24x_1 + 6x_2 + 19x_3$$

<u>Discussion</u>: Note that $x_1 = x_1(t)$, $x_2 = x_2(t)$, and $x_3 = x_3(t)$ are functions of t. Let $\mathbf{x}' = \frac{d\mathbf{x}}{dt}$, $\mathbf{x} = \begin{bmatrix} x_1 \\ x_2 \\ x_3 \end{bmatrix}$, and $A = \begin{bmatrix} -8 & 4 & 9 \\ 33 & -9 & -27 \\ -24 & 6 & 19 \end{bmatrix}$. We can then write the system of linear differential equations above as $\mathbf{x}' = A\mathbf{x}$. If A happened to be a diagonal matrix, $\begin{bmatrix} p & 0 & 0 \\ 0 & q & 0 \\ 0 & 0 & r \end{bmatrix}$, the system of equations would be

$$x_1'(t) = p\,x_1(t)$$
$$x_2'(t) = q\,x_2(t)$$
$$x_3'(t) = r\,x_3(t)$$

This system is called *uncoupled* because each equation involves a single variable that does not appear in the others. Therefore, each equation can be solved separately to yield

$$x_1(t) = a\,e^{pt}$$
$$x_2(t) = b\,e^{qt}$$
$$x_3(t) = c\,e^{rt}$$

Since A is not a diagonal matrix, it is not obvious how to solve the given system; but suppose A is diagonalizable; that is, $P^{-1}AP = D$, where D is a diagonal matrix. The idea is to make a substitution that will *uncouple* the system. Let $\mathbf{y} = \begin{bmatrix} y_1(t) \\ y_2(t) \\ y_3(t) \end{bmatrix}$ and put $\mathbf{x} = P\mathbf{y}$. Then

$$\mathbf{x}' = A\mathbf{x}$$
$$P\mathbf{y}' = AP\mathbf{y}$$
$$\mathbf{y}' = (P^{-1}AP)\mathbf{y}$$
$$\mathbf{y}' = D\mathbf{y}$$

This last system is uncoupled, and we can solve it as above. Once **y** has been found, we can recover **x** using **x** = P**y**.

<u>Solution</u>: Using the methods of Chapter 15, we find P so that $P^{-1}AP = D$ is diagonal. The diagonalization process will not be shown here. The matrix $P =$
$\begin{bmatrix} 1 & -1 & 3 \\ -3 & 4 & -9 \\ 2 & -3 & 7 \end{bmatrix}$ gives the diagonal matrix $P^{-1}AP = \begin{bmatrix} -2 & 0 & 0 \\ 0 & 3 & 0 \\ 0 & 0 & 1 \end{bmatrix}$. Solving (by inspection) the uncoupled system **y**$' = D$**y** yields

$$\begin{aligned} x_1(t) &= a\,e^{-2t} \\ x_2(t) &= b\,e^{3t} \\ x_3(t) &= c\,e^t \end{aligned}$$

Now, **x** = P**y** = $\begin{bmatrix} 1 & -1 & 3 \\ -3 & 4 & -9 \\ 2 & -3 & 7 \end{bmatrix} \begin{bmatrix} a\,e^{-2t} \\ b\,e^{3t} \\ c\,e^t \end{bmatrix} = \begin{bmatrix} ae^{-2t} - be^{3t} + 3ce^t \\ -3ae^{-2t} + 4be^{3t} - 9ce^t \\ 2ae^{-2t} - 3be^{3t} + 7ce^t \end{bmatrix}$.

This process is seen in Figure 18.1. Note the order in which the functions appear in line 3.

```
> P := matrix(3,3, [1,-1,3, -3,4,-9, 2,-3,7]):
> y_soln := vector([a*exp(-2*t),b*exp(3*t),c*exp(t)]);
```

$$y\_soln := \begin{bmatrix} a\,e^{(-2t)} & b\,e^{(3t)} & c\,e^t \end{bmatrix}$$

```
> x_soln := evalm(P &* y_soln);
```

$$x\_soln := \begin{bmatrix} a\,e^{(-2t)} - b\,e^{(3t)} + 3c\,e^t & -3a\,e^{(-2t)} + 4b\,e^{(3t)} - 9c\,e^t \\ 2a\,e^{(-2t)} - 3b\,e^{(3t)} + 7c\,e^t \end{bmatrix}$$

Figure 18.1: | The solution of the system of differential equations |

Once we have arrived at a final answer, we should check it as follows: To get the right-hand side of the original system, evaluate A &* x_soln as in line 1 of Figure 18.2. To get the left-hand side of the original system, map the derivative command onto x_soln. The result appears as line 2 in Figure 18.2. To finish checking, simply use Maple's equal test; cf. line 3 in Figure 18.2.

```
> A_x := evalm(A &* x_soln);
```

$$A\_x := \left[-2a\,e^{(-2t)} - 3b\,e^{(3t)} + 3c\,e^t \quad 6a\,e^{(-2t)} + 12b\,e^{(3t)} - 9c\,e^t \right.$$
$$\left. -4a\,e^{(-2t)} - 9b\,e^{(3t)} + 7c\,e^t \right]$$

```
> x_prime := map(diff, x_soln, t);
```

$$x\_prime := \left[-2a\,e^{(-2t)} - 3b\,e^{(3t)} + 3c\,e^t \quad 6a\,e^{(-2t)} + 12b\,e^{(3t)} - 9c\,e^t \right.$$
$$\left. -4a\,e^{(-2t)} - 9b\,e^{(3t)} + 7c\,e^t \right]$$

```
> equal(x_prime, A_x);
```

$$true$$

Figure 18.2: | Checking the solution of the system |

Solved Problem 2: | **This problem requires the LU program of Appendix B**.

Use the LU decomposition to solve the following system of equations.

$$
\begin{aligned}
2x_1 + 6x_2 + 4x_3 + 2x_4 &= 2 \\
x_1 + 6x_2 + 11x_3 + 7x_4 &= 4 \\
3x_1 + 11x_2 + 17x_3 + 12x_4 &= 10 \\
2x_1 + 10x_2 + 18x_3 + 13x_4 &= 9
\end{aligned}
$$

Discussion: Scientists and engineers often encounter systems of equations that are large enough to make computation time and round-off error significant issues. A number of methods have been developed to deal with this problem. We will look at two: LU (Lower-Upper) factorization in Solved Problem 2 and Gauss-Seidel iteration in Solved Problem 3.

Theorem: LU **Factorization**. If A is a matrix that can be put into echelon form without row interchanges, then there is a lower triangular matrix L and an upper triangular matrix U such that $A = LU$.

The code in the LU program of Appendix B will produce the required decomposition. It will be used to illustrate how the LU decomposition is employed to solve a system of equations.

<u>Solution</u>: Of course, Maple could solve this system directly, but we are asked to use the *LU* decomposition. The first step is to write the system in matrix form.

$$\begin{bmatrix} 2 & 6 & 4 & 2 \\ 1 & 6 & 11 & 7 \\ 3 & 11 & 17 & 12 \\ 2 & 10 & 18 & 13 \end{bmatrix} \begin{bmatrix} x_1 \\ x_2 \\ x_3 \\ x_4 \end{bmatrix} = \begin{bmatrix} 2 \\ 4 \\ 10 \\ 9 \end{bmatrix}$$

Enter the coefficient matrix and name it A. The next step is to obtain the LU factorization of A. To load the LU program of Appendix B, follow the appropriate steps, either 1 or 2, as indicated by your instructor:

1. Enter **read** 'LU.m'.

2. Open the file "LU Decomposition" as a scratchpad; select all of the LU function, and press $\boxed{enter}$ to define LU.

The program returns both the lower and upper matrices in a list. Set L and U to be the lower and upper matrices, respectively, as seen in line 1 of Figure 18.3. (The reader should use Maple to verify that the product *LU* does in fact equal A.)

```
> LU(A);
  L := ''[1]:
  U := ''[2]:
```

$$\left[\begin{bmatrix} 2 & 0 & 0 & 0 \\ 1 & 3 & 0 & 0 \\ 3 & 2 & 5 & 0 \\ 2 & 4 & 2 & 1 \end{bmatrix}, \begin{bmatrix} 1 & 3 & 2 & 1 \\ 0 & 1 & 3 & 2 \\ 0 & 0 & 1 & 1 \\ 0 & 0 & 0 & 1 \end{bmatrix} \right]$$

Figure 18.3: $\boxed{\text{The } LU \text{ decomposition of a matrix}}$

With this factorization, we may write our system of equations as follows.

$$\begin{bmatrix} 2 & 0 & 0 & 0 \\ 1 & 3 & 0 & 0 \\ 3 & 2 & 5 & 0 \\ 2 & 4 & 2 & 1 \end{bmatrix} \left(\begin{bmatrix} 1 & 3 & 2 & 1 \\ 0 & 1 & 3 & 2 \\ 0 & 0 & 1 & 1 \\ 0 & 0 & 0 & 1 \end{bmatrix} \begin{bmatrix} x_1 \\ x_2 \\ x_3 \\ x_4 \end{bmatrix} \right) = \begin{bmatrix} 2 \\ 4 \\ 10 \\ 9 \end{bmatrix}$$

Introducing new variables, we can replace this system of equations by two simpler systems of equations.

$$\begin{bmatrix} 2 & 0 & 0 & 0 \\ 1 & 3 & 0 & 0 \\ 3 & 2 & 5 & 0 \\ 2 & 4 & 2 & 1 \end{bmatrix} \begin{bmatrix} y_1 \\ y_2 \\ y_3 \\ y_4 \end{bmatrix} = \begin{bmatrix} 2 \\ 4 \\ 10 \\ 9 \end{bmatrix} \tag{18.1}$$

$$\begin{bmatrix} 1 & 3 & 2 & 1 \\ 0 & 1 & 3 & 2 \\ 0 & 0 & 1 & 1 \\ 0 & 0 & 0 & 1 \end{bmatrix} \begin{bmatrix} x_1 \\ x_2 \\ x_3 \\ x_4 \end{bmatrix} = \begin{bmatrix} y_1 \\ y_2 \\ y_3 \\ y_4 \end{bmatrix} \tag{18.2}$$

Rewrite (18.1) as a system of equations:

$$\begin{aligned} 2y_1 &= 2 \\ y_1 + 3y_2 &= 4 \\ 3y_1 + 2y_2 + 5y_3 &= 10 \\ 2y_1 + 4y_2 + 2y_3 + y_4 &= 9 \end{aligned}$$

This system is easily solved by forward substitution to yield $y_1 = 1$, $y_2 = 1$, $y_3 = 1$, $y_4 = 1$. Using this solution, we write (18.2) as a system of equations.

$$\begin{aligned} x_1 + 3x_2 + 2x_3 + x_4 &= 1 \\ x_2 + 3x_3 + 2x_4 &= 1 \\ x_3 + x_4 &= 1 \\ x_4 &= 1 \end{aligned}$$

Solve this system by back substitution to yield: $x_1 = 4$, $x_2 = -1$, $x_3 = 0$, $x_4 = 1$. (The reader should check the answers with Maple.)

For a small system of equations, this is not an efficient method, but for a large system, the LU factorization allows a complicated system of equations to be replaced by two systems of equations that can be easily solved by forward and backward substitution. Very efficient computer code exists to implement these procedures.

Solved Problem 3: Gauss-Seidel iteration. Let $A = \begin{bmatrix} 5 & 2 & 1 & 1 \\ 3 & 6 & 3 & 1 \\ 2 & 2 & 7 & 3 \\ 1 & 2 & 3 & 6 \end{bmatrix}$ and $\mathbf{b} = $

$(4, 3, 2, 1)$. Use ten iterations of the Gauss-Seidel method to approximate the solution of the system of equations $A\mathbf{x} = \mathbf{b}$.

Discussion: In some cases, it is better to approximate the solution of a system of equations rather than attempt to solve it exactly. The Gauss-Seidel method is such an approximation technique. It is a refinement of a similar scheme called the Jacobi method, about which we'll say a bit more later.

For this system of equations, the exact solution in line 3 of Figure 18.4 is easily obtained using Maple (with the `matsolve` function of Appendix B). The sole purpose of this problem is to illustrate how the Gauss-Seidel method works. What we will present is a simplified version of the actual procedure used in practice.

Solution: Let M denote the lower triangular part of A; that is,

$$M = \begin{bmatrix} 5 & 0 & 0 & 0 \\ 3 & 6 & 0 & 0 \\ 2 & 2 & 7 & 0 \\ 1 & 2 & 3 & 6 \end{bmatrix}$$

Then, the original system of equations can be written as $M\mathbf{x} = (M - A)\mathbf{x} + \mathbf{b}$, or $\mathbf{x} = (I - M^{-1}A)\mathbf{x} + M^{-1}\mathbf{b}$. The idea is to begin with a guess at the solution (even a bad one such as $\mathbf{x}_0 = (0,0,0,0)$) and improve it using the formula

$$\mathbf{x}_{n+1} = (I - M^{-1}A)\mathbf{x}_n + M^{-1}\mathbf{b}$$

Enter A, $\mathbf{x}$, $\mathbf{b}$, and M, and, to simplify typing, define `Mi` and `M1` as in Figure 18.4.

```
> A := matrix(4,4, [5,2,1,1, 3,6,3,1, 2,2,7,3, 1,2,3,6]):
  x := vector([x1,x2,x3,x4]):    b := vector([4,3,2,1]):
> M := matrix(4,4, [5,0,0,0, 3,6,0,0, 2,2,7,0, 1,2,3,6]):
  Mi := M^(-1):    M1:=evalm(Id - Mi &* A):
> matsolve(A &* x = b);
```

$$\left\{ x1 = \frac{131}{174},\ x2 = \frac{71}{696},\ x3 = \frac{17}{348},\ x4 = \frac{-1}{58} \right\}$$

```
> evalf(");
```

$$\{x1 = .7528735632,\ x2 = .1020114943,\ x3 = .04885057471,\ x4 = -.01724137931\}$$

Figure 18.4: Preparing for Gauss-Seidel iteration

Now enter the function `GS := x -> evalf(evalm(M1 &* x + Mi &* b))` as seen in Figure 18.5. The function `GS` is decimal approximation composed with one step of Gauss-Seidel iteration: $\mathbf{x}_{n+1} := (I - M^{-1}A)\mathbf{x}_n + M^{-1}\mathbf{b}$.

The value of `GS(x)` is plugged back into the formula to get a third value, `GS(GS(x))`. This procedure is repeated 10 times. To shorten the work, use Maple's `seq` function with repeated composition via the `@@` operator. Repeated composition is done as in `(GS@@3)(x)`, which is Maple's notation for $(GS \circ GS \circ GS)(x)$ which is equal to $GS(GS(GS(x)))$. We combine the results of lines 2 and 3 of Figure 18.5 to get the matrix in line 4. The rows of this matrix are successive approximations to a solution of the system. Compare the last row of the matrix in line 4 of Figure 18.5 with the approximate solution of the system shown in line 4 of Figure 18.4.

```
> GS := x -> evalf(evalm(M1 &* x + Mi &* b));
```

$$GS := x \rightarrow \text{evalf}\left(\text{evalm}\left((M1 \ \&^* \ x) + (Mi \ \&^* \ b)\right)\right)$$

```
> GS(vector([0,0,0,0]));
```

$$[.800000 \quad .100000 \quad .0285714 \quad -.0142857]$$

```
> seq((GS@@i)("),i=1..9):
> matrix([[0,0,0,0]," ","]);
```

$$
\begin{bmatrix}
0 & 0 & 0 & 0 \\
.800000000 & .1000000000 & .02857142857 & -.01428571429 \\
.7571428572 & .1095238095 & .04421768708 & -.01814058957 \\
.7509750567 & .1054270597 & .04880250513 & -.01803944859 \\
.7516765648 & .1027670398 & .04931873380 & -.01752780763 \\
.7525349989 & .1019944349 & .04907493647 & -.01729144636 \\
.7528455280 & .1019216755 & .04890570458 & -.01723433213 \\
.7528970553 & .1019710087 & .04885240975 & -.01723271701 \\
.7528876580 & .1020020856 & .04884552340 & -.01723806658 \\
.7528876744 & .1020114122 & .04884800379 & -.01724075170 \\
.7528739847 & .1020124644 & .04884990813 & -.01724143965
\end{bmatrix}
$$

Figure 18.5: | Ten iterations of the Gauss-Seidel method |

Remark on the Jacobi method: It is easy to see that our derivation of the formula $\mathbf{x} = (I - M^{-1}A)\mathbf{x} + M^{-1}\mathbf{b}$ from $A\mathbf{x} = \mathbf{b}$ holds if M is *any* invertible matrix! The Jacobi method is the same procedure we just executed with M replaced by the diagonal of A:

$$\begin{bmatrix} 5 & 0 & 0 & 0 \\ 0 & 6 & 0 & 0 \\ 0 & 0 & 7 & 0 \\ 0 & 0 & 0 & 6 \end{bmatrix}$$

It is a bit less efficient than the Gauss-Seidel method.

Solved Problem 4: The generalized inverse and curve fitting. In an experiment the following data were collected:

x	1	1.5	2	2.5	3
y	3.4	3.8	3.9	4.3	4.5

(a) Find the least squares straight line fit to this data and plot both the line and the data.

(b) Find the best least squares fit to this data by a function of the form $ax + b\sin(x) + c\cos(x)$. Calculate the (approximate) generalized inverse of the associated matrix and plot both the function and the data.

Discussion: If M is a matrix (not necessarily square) whose columns are linearly independent, then it can be shown that M^tM is invertible. (See Exploration and Discovery Problem 2). When this is the case, the matrix $(M^tM)^{-1}M^t$ is called the *generalized* (or *pseudo-*) inverse of M.

If we want to solve a system of equations $M\mathbf{x} = \mathbf{y}$ and there are more equations than unknowns, there may not be a solution, so we look for the nearest thing to one — literally. To say the system $M\mathbf{x} = \mathbf{y}$ has no solution means that the vector $\mathbf{y}$ does not belong to the set of all vectors of the form $M\mathbf{x}$ (the column space of M); however, there *is* a unique vector in the column space that is nearest to it. That vector is $(M^tM)^{-1}M^t\mathbf{y}$.

Solution to (a): Ideally, we want to find a and b for which the line $a + bx$ goes through the data. That is,

$$a + b(1) \quad = \quad 3.4$$
$$a + b(1.5) \quad = \quad 3.8$$
$$a + b(2) \quad = \quad 3.9$$
$$a + b(2.5) \quad = \quad 4.3$$
$$a + b(3) \quad = \quad 4.5$$

This means we want to solve $M \begin{bmatrix} a \\ b \end{bmatrix} = \mathbf{y}$, where $M = \begin{bmatrix} 1 & 1 \\ 1 & 1.5 \\ 1 & 2 \\ 1 & 2.5 \\ 1 & 3 \end{bmatrix}$ and $\mathbf{y} = \begin{bmatrix} 3.4 \\ 3.8 \\ 3.9 \\ 4.3 \\ 4.5 \end{bmatrix}$,

but there is no solution. (Why?) Thus, we find the best approximation with the generalized inverse $(M^t M)^{-1} M^t \mathbf{y}$. The line we seek then is $a + bx$ where

$$\begin{bmatrix} a \\ b \end{bmatrix} = (M^t M)^{-1} M^t \mathbf{y}$$

Enter each of the vectors

$$x\_data := (1.0, 1.5, 2.0, 2.5, 3.0)$$
$$y\_data := (3.4, 3.8, 3.9, 4.3, 4.5)$$

as in Figure 18.6. (For convenience, we enter the data as lists rather than vectors.) Define the matrix M and M^t with

```
M := augment([1,1,1,1,1], x_data, y_data)
Mt := transpose(M)
```

Maple Remark: Entering matrices of data can be a chore. If your data have been generated by another program or if you have entered the data in a file, it can be read with the importdata command from the stats package. (Also see ?readline and ?sscanf.)

Next enter and evaluate as a matrix (Mt &* M)^(-1)&* Mt &* y_data. The resulting vector is in line 2 of Figure 18.6. The least squares fit is shown in line 3 of Figure 18.6.

Now we must plot the line along with the data points. (See Section A.12 in Appendix A for details on plotting. If you use a version of Maple earlier than release 3, omit the symbol portion of the plot statement.)

```
>  x_data := [1.0, 1.5, 2.0, 2.5, 3.0]:
   y_data := [3.4, 3.8, 3.9, 4.3, 4.5]:
   M := augment([1,1,1,1,1], x_data):
   Mt := transpose(M):
>  evalm((Mt &* M)^(-1)&* Mt &* y_data);
```

$$[2.900000000 \quad .540000000]$$

```
>  my_line := dotprod(",vector([1,x]));
```

$$my\_line := 2.900000000 + .540000000\,x$$

```
>  line_plot := plot(my_line,x=0..4,y=0..5):
   data_plot := plot(augment(x_data,y_data),x=0..4,y=0..5,
     style=point, symbol=diamond):
   plots[display]([line_plot,data_plot]);
```

Figure 18.6: Fitting a line to data.

The `leastsquare` curve-fitting function: Maple automates the process we just went through with the built-in function `leastsquare` from the `fit` "sub-" package of the `stats` package. The data may be entered with the x-coordinates in the first vector and the y-coordinates in the second. Notice that the variables are given as a vector in an index to the function. You can see the syntax in Figure 18.7—pay close attention to brackets and parentheses. In more general cases, the form of the function you want to fit is entered along with the variables in equation form, followed by a set made of the constants to determine. For example, if you need to fit a general quadratic to the data, you would enter the index as `[[x,y], y=a*x^2+b*x+c, {a,b,c}]`. You may fit any linear combination of functions.

```
> with(stats):
  with(fit):
> leastsquare[[x,y]]([x_data, y_data]);
```

$$y = 2.900000000 + .540000000\,x$$

Figure 18.7: | The fit function for a line |

Solution to (b): Ideally, we want to find a and b and c for which $a\sin(x)+b\cos(x)+cx$ goes through the data. That is,

$$
\begin{aligned}
a\sin(1.0) + b\cos(1.0) + c &= 3.4 \\
a\sin(1.5) + b\cos(1.5) + 1.5c &= 3.8 \\
a\sin(2.0) + b\cos(2.0) + 2c &= 3.9 \\
a\sin(2.5) + b\cos(2.5) + 2.5c &= 4.3 \\
a\sin(3.0) + b\cos(3.0) + 3c &= 4.5
\end{aligned}
$$

However, there is no solution to this system, so we must find the generalized inverse of the coefficient matrix to get a "best fit." We need to generate the vector

$$[\sin(1.0), \sin(1.5), \sin(2.0), \sin(2.5), \sin(3.0)]$$

of b's coefficients. This is done most efficiently with `map` by `map(sin,x_data)` as seen in line 1 of Figure 18.8. Do the same for the cosines (c's coefficients). We enter `augment(x_data, b_coeffs, c_coeffs)` to define the matrix M.

Next calculate M's generalized inverse; it appears in line 2 in Figure 18.8. Now evaluate `Mg &* y_data`. The vector in line 3 gives the coefficients of x, $\sin(x)$, and $\cos(x)$ in order. Using `dotprod` gives a convenient way to obtain the equation.

```
> b_coeffs := map(sin,x_data):   c_coeffs := map(cos,x_data):
  M := augment(x_data, b_coeffs, c_coeffs):
> Mg := (transpose(M) &* M)^(-1) &* transpose(M);
```

$$Mg := \begin{bmatrix} .5500520546 & -.0874760212 & -.3895479802 & -.177526430 & .601358005 \\ -.620374263 & .5151379496 & .973978354 & .460290794 & -1.083672124 \\ 1.474565008 & -.0655802326 & -.939590143 & -.716783614 & .764981546 \end{bmatrix}$$

```
> evalm(Mg &* y_data);
```

$$[1.961278356 \quad .749493151 \quad 1.460162002]$$

```
> my_fit := dotprod('', [x,sin(x),cos(x)]);
```

$$my\_fit := 1.961278356\,x + .749493151\,\sin(x) + 1.460162002\,\cos(x)$$

Figure 18.8: | Fitting $a\,x + b\sin(x) + c\cos(x)$ to data |

In Figure 18.9, we use the `leastsquare` fit function. The syntax of `leastsquare` requires us to use the index `[[x,y],eq,{a,b,c}]` where eq is $y = a\sin(x) + b\cos(x) + cx$. Check to see how this result compares with that of Figure 18.8. The plots of the data and the function appear in Figure 18.10.

```
> eq := y = a*x + b*sin(x) + c*cos(x);
```

$$eq := y = a\,x + b\,\sin(x) + c\,\cos(x)$$

```
> leastsquare[[x,y],eq,{a,b,c}]([x_data, y_data]);
```

$$y = 1.961278361\,x + .7494931795\,\sin(x) + 1.460162061\,\cos(x)$$

Figure 18.9: | The generalized inverse and fit for $a\,x + b\sin(x) + c\cos(x)$ |

```
> fit_plot := plot(my_fit,x=0..4,y=0..5):
  data_plot := plot(augment(x_data,y_data),x=0..4,y=0..5,
    style=point, symbol=diamond):
  plots[display]([fit_plot,data_plot]);
```

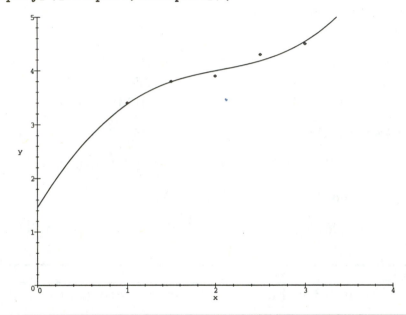

Figure 18.10: $\boxed{\text{Graph of } a\,x + b\sin(x) + c\cos(x) \text{ and data}}$

Solved Problem 5: Plot the graph of $4x^2 + 3xy + y^2 + 2x + y - 1 = 0$. Apply a rotation through 22.5 degrees and plot the graph again.

Solution: The implicit plotting routines are part of the **plots** package, so we must first load them via **with(plots)**. Enter the definition

$$\text{eq := 4*x\^{}2 + 3*x*y + y\^{}2 + 2*x + y - 1 = 0}$$

and then use **implicitplot** as in line 2 of Figure 18.11.

```
> eq := 4*x^2 + 3*x*y + y^2 + 2*x + y - 1 = 0:
> implicitplot(eq, x=-3..3, y=-3..3);
```

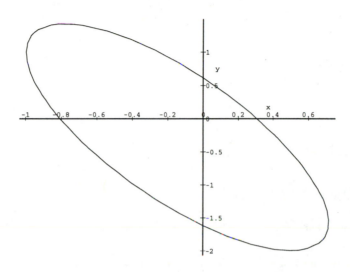

Figure 18.11: Ellipse

Next we enter the rotation matrix

$$M\_rot := \begin{bmatrix} \cos(22.5°) & -\sin(22.5°) \\ \sin(22.5°) & \cos(22.5°) \end{bmatrix}$$

as in line 1 of Figure 18.12. (Remember to enter $22.5°$ as $\pi/8$ radians.) To get the rotation transformation, find $M\_rot\,\mathbf{x}$. We obtain the result

$$\left[\frac{1}{2}\sqrt{2+\sqrt{2}}\,x - \frac{1}{2}\sqrt{2-\sqrt{2}}\,y \quad \frac{1}{2}\sqrt{2-\sqrt{2}}\,x + \frac{1}{2}\sqrt{2+\sqrt{2}}\,y\right]$$

as line 2.

Now substitute the first coordinate of the result in line 2 of Figure 18.12 for x in

$$4x^2 + 3xy + y^2 + 2x + y - 1 = 0$$

and the second coordinate for y. Before plotting (Figure 18.13), we will apply several (nested) simplifying commands to the result.

```
> M_rot:=matrix(2,2, [cos(Pi/8),-sin(Pi/8), cos(Pi/8),sin(Pi/8)]);
```

$$M\_rot := \begin{bmatrix} \frac{1}{2}\sqrt{2+\sqrt{2}} & -\frac{1}{2}\sqrt{2-\sqrt{2}} \\ \frac{1}{2}\sqrt{2-\sqrt{2}} & \frac{1}{2}\sqrt{2+\sqrt{2}} \end{bmatrix}$$

```
> evalm(M_rot&*[x,y]);
```

$$\left[\frac{1}{2}\sqrt{2+\sqrt{2}}\,x - \frac{1}{2}\sqrt{2-\sqrt{2}}\,y \quad \frac{1}{2}\sqrt{2-\sqrt{2}}\,x + \frac{1}{2}\sqrt{2+\sqrt{2}}\,y \right]$$

```
> subs({x=''[1],y=''[2]}, eq);
```

$$4\left(\frac{1}{2}\sqrt{2+\sqrt{2}}\,x - \frac{1}{2}\sqrt{2-\sqrt{2}}\,y\right)^2$$
$$+3\left(\frac{1}{2}\sqrt{2+\sqrt{2}}\,x - \frac{1}{2}\sqrt{2-\sqrt{2}}\,y\right)\left(\frac{1}{2}\sqrt{2-\sqrt{2}}\,x + \frac{1}{2}\sqrt{2+\sqrt{2}}\,y\right)$$
$$+\left(\frac{1}{2}\sqrt{2-\sqrt{2}}\,x + \frac{1}{2}\sqrt{2+\sqrt{2}}\,y\right)^2 + \sqrt{2+\sqrt{2}}\,x - \sqrt{2-\sqrt{2}}\,y + \frac{1}{2}\sqrt{2-\sqrt{2}}\,x$$
$$+\frac{1}{2}\sqrt{2+\sqrt{2}}\,y - 1 = 0$$

```
> rot_eq := collect(combine(expand(''),radical),{x,y});
```

$$\left(\frac{5}{2} - \frac{3}{2}\sqrt{2}\right)y^2 + \left(\frac{1}{2}\sqrt{2+\sqrt{2}} - \sqrt{2-\sqrt{2}}\right)y + \left(\frac{5}{2} + \frac{3}{2}\sqrt{2}\right)x^2$$
$$+ \left(\frac{1}{2}\sqrt{2-\sqrt{2}} + \sqrt{2+\sqrt{2}}\right)x - 1 = 0$$

Figure 18.12: $\boxed{\text{Rotated Ellipse, I}}$

Looking at line 4, Figure 18.12, we see the simplifications, from the inside out, are:

1. `expand`:   multiply the products,
2. `combine` ... `radical`:   simplify and combine like radicals,
3. `collect` ... `{x,y}`:   collect like terms in the variables **x** and **y**.

```
>  implicitplot({eq, rot_eq} ,x=-3..3, y=-3..3,
     view=[-1.5..1.5, -2..2], scaling=constrained);
```

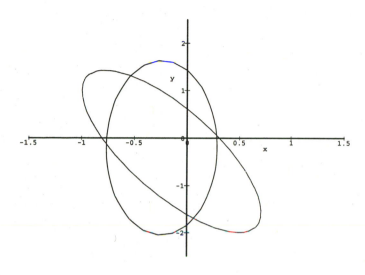

Figure 18.13: Rotated Ellipse, II

Solved Problem 6: This exercise uses the program QR in Appendix B.

Let $A = \begin{bmatrix} 1 & 3 & 5 \\ 2 & 3 & 1 \\ 2 & 4 & -1 \end{bmatrix}$. Find the QR factorization of A. Calculate the eigenvalues of A and compare them with approximations obtained by executing 5, 10, and 30 steps of the QR algorithm.

Discussion: It is not practical to find eigenvalues of large matrices by finding the roots of its characteristic polynomial. A procedure known as the QR algorithm efficiently approximates the eigenvalues in many cases. The algorithm begins with the QR factorization theorem.

Theorem: QR factorization. Let A be a nonsingular square matrix. Then there is an orthogonal matrix Q and a nonsingular upper triangular matrix R such that $A = QR$.

The idea behind the QR factorization is really quite simple. The columns of the matrix Q are just the result of applying the Gram-Schmidt algorithm to the columns of A; the entries of the matrix R contain the information necessary to keep track of the operations required to apply the algorithm. There are many applications of the QR factorization, but among the most important is the following procedure known as the QR algorithm. This is a simplified version of what is used in practice.

The QR Algorithm

1. If A is a square nonsingular matrix, obtain the QR factorization, $A = QR$, and let $A_1 = RQ$ (the product of the QR factorization in reverse order).

2. Continue the process:

   A_2 is the QR factorization of A_1 taken in reverse order, and so on.

   In general, A_n is the result of applying this procedure n times.

For a matrix with real eigenvalues, some fairly general conditions will ensure that the QR algorithm converges rapidly to a matrix that is *approximately upper triangular* (i.e., the entries below the main diagonal have very small magnitude) and the entries on the main diagonal are approximations of the eigenvalues of A.

In the QR Decomposition section of Appendix B, the Maple code provided calculates the QR factorization (QR) and executes the QR algorithm (QR_algorithm). We use this code in the following solution.

Solution: First enter the program QR (either by keying it in or through a Maple scratchpad). Enter the matrix and name it A. Set Q equal to first element of QR(A) and R to the second.. The results are seen in lines 1 and 2 of Figure 18.14.

```
> A := matrix(3,3, [1,3,5, 2,3,1, 2,4,-1]):
> QR(A);
  Q := ''[1]:
  R := '''''[2]:
```

$$\left[\begin{bmatrix} \frac{1}{3} & \frac{10}{51}\sqrt{17} & \frac{2}{17}\sqrt{17} \\ \frac{2}{3} & -\frac{7}{51}\sqrt{17} & \frac{2}{17}\sqrt{17} \\ \frac{2}{3} & \frac{2}{51}\sqrt{17} & -\frac{3}{17}\sqrt{17} \end{bmatrix} , \begin{bmatrix} 3 & \frac{17}{3} & \frac{5}{3} \\ 0 & \frac{1}{3}\sqrt{17} & \frac{41}{51}\sqrt{17} \\ 0 & 0 & \frac{15}{17}\sqrt{17} \end{bmatrix} \right]$$

Figure 18.14: The QR factorization of A

The matrix R is obviously upper triangular. The reader should use Maple to verify at least 1 and 3 below:

1. Q is an orthogonal matrix.

2. The columns of Q are the result of applying the Gram-Schmidt algorithm to the columns of A.

3. $QR = A$.

Approximating the eigenvalues of A with `fsolve` yields -2.708563873, -0.8450309055, and 6.553594778.

Even with Maple to help, it is tedious to execute the individual steps of the QR algorithm. The function `QR_algorithm(A,n)` provided in Appendix B automates the procedure. Calculate `QR_algorithm(A,5)`. The result appears in line 1 of Figure 18.15. Observe that the entries below the main diagonal are relatively small in magnitude and that the entries on the main diagonal are close to the eigenvalues we calculated earlier. The command `fnormal` inside the program forces very small numbers to zero. Notice also in lines 2 and 3 of Figure 18.15 that the approximations become much better as the number of iterations increases. A more refined version of this algorithm will converge more rapidly.

```
> QR_algorithm(A,5);
```

$$\begin{bmatrix} 6.57645483 & 2.295408477 & -1.697610320 \\ -.09153286161 & -2.769777629 & -3.204392021 \\ .001193701151 & .02308007801 & -.8066788502 \end{bmatrix}$$

```
> QR_algorithm(A,10);
```

$$\begin{bmatrix} 6.553309736 & -2.408194533 & 1.637885379 \\ -.001096216092 & -2.708391621 & -3.243884859 \\ .4157274527\ 10^{-7} & .00006478880359 & -.8449181144 \end{bmatrix}$$

```
> QR_algorithm(A,30);
```

$$\begin{bmatrix} 6.553594766 & -2.407155259 & 1.638185597 \\ 0 & -2.708563875 & -3.243755761 \\ 0 & 0 & -.8450308998 \end{bmatrix}$$

Figure 18.15: The QR algorithm applied to A

18.3   Exercises

1. Solve the following system of linear differential equations and check your solution.

$$\begin{aligned}
x_1' &= 11x_1 + 4x_2 + 4x_3 + 12x_4 \\
x_2' &= 16x_1 + 3x_2 + 8x_3 + 16x_4 \\
x_3' &= 8x_1 + 7x_3 + 8x_4 \\
x_4' &= -16x_1 - 4x_2 - 8x_3 - 17x_4
\end{aligned}$$

2. Use 3, 7, and 10 iterations of the Gauss-Seidel method to approximate the solution of the following system of equations. Compare your answer with the exact solution in each case.

$$\begin{aligned}
6x + 2y - z + w &= 4 \\
2x + 7y + z - w &= 5 \\
3x - y + 5z + 2w &= 1 \\
4x + 3y + 2z - 8w &= 2
\end{aligned}$$

3. Use the LU factorization to solve the system of equations in Exercise 2. Show all the steps in your solution.

4. Show that if M has an inverse, then $(M^t M)^{-1} M^t = M^{-1}$. (No computer is needed for this exercise.)

5. Find the best fit to the data below by a function of the form $ax + b\sin(x) + c\cos(x)$. Show the matrix M and its generalized inverse. Plot the function and data points.

x	3	3.5	4	4.5	5	5.5	6	6.5	7	7.5	8
y	1.3	1.4	1.6	1.5	1.5	1.4	1.5	1.5	1.4	1.2	1.3

6. Find the best fit to the data above by a function of the form $ax + b\ln(x)$. Show the matrix M and its generalized inverse. Plot the function and data points.

7. Repeat Solved Problem 5 for $4x^2 - xy + y^2 + 2x + y - 1 = 0$.

8. Repeat Solved Problem 5 for $4x^2 - xy + y^2 + 2x + y - 1 = 0$ using an angle of $60°$.

18.4   Exploration and Discovery

1. Implement the Jacobi method discussed in Solved Problem 3 and apply it to Exercise 2. Compare the Jacobi and Gauss-Seidel methods step by step for several iterations in this problem and discuss your observations. Compare the methods for a 3 by 3 system.

2. This is not a computer problem. It's all pencil and paper. Suppose M is m by n and has linearly independent columns. Prove that $M^t M$ has an inverse by the following steps:

   (a) Show that $M^t M$ is square, and therefore it suffices to show that if $(M^t M)\mathbf{x} = \mathbf{0}$, then $\mathbf{x} = \mathbf{0}$.

   (b) Show that if $\mathbf{u}$ is in $\mathbb{R}^m$ and $\mathbf{v}$ is in $\mathbb{R}^n$ then $(M^t \mathbf{u}) \cdot \mathbf{v} = \mathbf{u} \cdot (M\mathbf{v})$.

   (c) Let $\mathbf{x}$ be in $\mathbb{R}^n$ and substitute $\mathbf{u} = M\mathbf{x}$ and $\mathbf{v} = \mathbf{x}$ in (b) to conclude that $|M\mathbf{x}|^2 = 0$.

   (d) Use (c) and the fact that M has linearly independent columns to conclude that $\mathbf{x} = \mathbf{0}$ as desired.

3. (a) Do Exercise 1.

   (b) Do Exploration and Discovery Problem 3(d) in Chapter 3.

   (c) Do Exploration and Discovery Problem 1(d) in Chapter 15 for e^x.

   (d) Let $\mathbf{c}$ be the vector of constants (c_1, c_2, c_3, c_4). Show that the solution of the system in Exercise 1 can be written as $\mathbf{x} = e^{At}\mathbf{c}$, where A is the coefficient matrix.

4. Rotating $ax^2 + bxy + cy^2 + dx + ey + k = 0$ through the proper angle will eliminate the xy term. Let's find such an angle.

   (a) Enter the rotation matrix $\begin{bmatrix} \cos(\theta) & -\sin(\theta) \\ \sin(\theta) & \cos(\theta) \end{bmatrix}$ and the quadratic $ax^2 + bxy + cy^2 + dx + ey + k$.

   (b) The linear transformation

   $$T(x, y) = \begin{bmatrix} \cos(\theta) & -\sin(\theta) \\ \sin(\theta) & \cos(\theta) \end{bmatrix} \begin{bmatrix} x \\ y \end{bmatrix}$$

rotates the vector (x, y) through the angle θ. Use subs to substitute the first coordinate of $T(x, y)$ into $ax^2 + bxy + cy^2 + dx + ey + k$ for x and the second for y.

(c) After you simplify and collect like terms in the preceding substitution, use coeff(coeff('',x),y) to extract all the terms with a factor of xy. (Be sure you get them all.)

(d) Now, set the coefficient of xy equal to zero and solve for θ. Discuss your observations and conclusions.

(e) Plot $4x^2 - xy + y^2 + 2x + y - 1 = 0$ as we did in Solved Problem 5.

(f) Use your result from (d) to find an angle through which to rotate

$$4x^2 - xy + y^2 + 2x + y - 1 = 0$$

to eliminate the xy term.

(g) Plot the rotated curve above as we did in Solved Problem 5.

18.5 Curve Fitting

LABORATORY EXERCISE 18–1

Name _____   Due Date _____

A small cannon is fired at a wall. The distance from the cannon to the wall, x meters, and the height of the spot on the wall where the projectile strikes, y meters, are recorded in the table:

x	3	3.5	4	4.5	5	5.5	6	6.5	7	7.5	8
y	1.3	1.4	1.6	1.5	1.5	1.4	1.5	1.5	1.4	1.2	1.3

1. We suspect that a quadratic $ax^2 + bx + c$ fits these data. Write down the matrix M whose generalized inverse must be found to do this.

2. Find the generalized inverse of the matrix M in 1.

3. Use the generalized inverse of the matrix M from 1 to find the best least squares quadratic $ax^2 + bx + c$ that fits these data.

4. Check your result by using Maple's `leastsquare` fitting function.

5. Plot the data points and the quadratic you have found. Comment on your findings. Do they seem reasonable?

6. We might surmise from the graph that our cannon was not actually on the ground. Suppose that we *know* it's on the ground and therefore we really suspect $ax^2 + bx$ is the right formula. Repeat all the parts above in this problem to find and plot the best least squares quadratic $ax^2 + bx$ for the data.

18.6   *QR* **Factorization**

LABORATORY EXERCISE 18–2

Name _____   Due Date _____

Let $A = \begin{bmatrix} 2 & 2 & 1 & -2 \\ 2 & 2 & 0 & -2 \\ -6 & 5 & 5 & 13 \\ 4 & -3 & -1 & -7 \end{bmatrix}$.

1. Find the *QR* factorization, $A = QR$.

2. Show that the columns of Q are the result of applying the Gram-Schmidt algorithm to the columns of A. (See Chapter 12.)

3. Use the characteristic polynomial to obtain and approximate the eigenvalues of A.

4. Use 5, 10, and 30 iterations of the *QR* algorithm to approximate the eigenvalues of A and compare the results with your answer in 3.

5. Show that if A_n is the result of applying n iterations of the *QR* algorithm to A, then A is similar to A_n and hence they have the same eigenvalues. (<u>Hint</u>: No computer is needed here. Show first that for nonsingular matrices B and C, BC is similar to CB.)

Appendix A
MAPLE V MINI-REFERENCE

This appendix is not intended as a substitute for the Maple manuals; owners of Maple will have *First Leaves* or the *Maple V Flight Manual*, lab users should have access through the lab's reference materials. When this book was written, the authors used Maple V Release 3. If you are using an earlier version, you will notice a few differences. We have tried to mention them on the rare occasions when they might cause confusion. The most conspicuous new feature is "real math" display.

The Maple program has a fine set of manuals, nevertheless many people read software documentation only as a last resort. This appendix is a concise reference to the major features of Maple that often arise in the exercises in this book. Each topic in this appendix refers the reader to *First Leaves* (FL) and *Flight Manual* (FM) by section for detailed information. The definitive reference for any Maple command is the *Maple Library Reference Manual*. The best comprehensive, yet concise, guide to Maple is Nancy Blachman and Michael Mossighoff's *Maple V Quick Reference*. For further study of computer mathematics systems, the best introductory book available is André Heck's *Introduction to Maple*. We encourage the interested reader to explore the Maple manuals and the *online help system*. (See Section A.14 for bibliographic details.)

The Internet provides a forum in which both specific and general questions about computer mathematics systems are answered and discussed. All that you need is access through a news feed. If you have an account on a university system try

"news" on VMS Systems
"nn" or "rn" on Unix Systems

then issue a "go sci.math.symbolic" command.

Waterloo Maple Software has technical support available through e-mail. For information about this, or any other aspect of Maple, send a message to:

info@maplesoft.on.ca

A.1 Special Functions and Symbols

Maple has predefined constants and symbols as follows:

Special Constants and Symbols

E	the number e
Pi	the number π
I	the imaginary number i or $\sqrt{-1}$
infinity	∞
-infinity	$-\infty$
&*()	the identity matrix I_n
''	the previous result

Entering arithmetic expressions involves customary syntax: addition (+), subtraction (-), division (/), exponentiation (^), and multiplication (*). Every multiplication, just as on a calculator, requires a *; 2*x is $2x$, while 2x gives an error, and x*2 gives $2x$, while x2 is the name $x2$ (some use this as a substitute for x_2).

Parentheses are used for grouping and functions, brackets are used for lists (especially in matrices and vectors), and braces are used for sets.

To cancel a command or to get out of an undesirable situation, use:

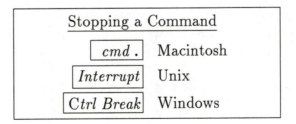

Stopping a Command

$\boxed{cmd \, .}$	Macintosh
$\boxed{Interrupt}$	Unix
$\boxed{Ctrl \; Break}$	Windows

A.2 Tutorial for Beginners

FL: 1.1 to 1.12   FM: Chapter 1 and 2.1

To enter an expression into Maple, type after the prompt symbol, usually a ">", end the line with a semicolon, and then press $\boxed{enter}$ to have Maple execute your directives. Type

$$x^2+3*x-4;$$

at the prompt and press $\boxed{enter}$ when done; we have entered $x^2 + 3x - 4$. In almost all instances, spaces can be added to improve readability. Type factor(''); then

$\boxed{enter}$, and observe that the factored form $(x+4)(x-1)$ appears in line 2. We will refer to a Maple input-ouput pair as a *line*.

```
> x^2 + 3*x - 4;
```
$$x^2 + 3x - 4$$

```
> factor(");
```
$$(x+4)(x-1)$$

```
> solve(x^2 + 3*x - 4);
```
$$1, -4$$

```
> subs(x=2, x^2 + 3*x - 4);
```
$$6$$

```
> a := 2;
  a^2;
```
$$a := 2$$
$$4$$

```
> b := 2*a - a^3;
```
$$b := -4$$

```
> a := 'a';
```
$$a := a$$

```
> y := x^2 + 3*x - 4;
  y^6;
```
$$y := x^2 + 3x - 4$$
$$\left(x^2 + 3x - 4\right)^6$$

```
> expand(");
```
$$28416\,x^2 - 18432\,x - 11280\,x^4 + x^{12} + 8568\,x^5 + 18\,x^{11} + 111\,x^{10} + 180\,x^9$$
$$+4096 - 705\,x^8 - 2142\,x^7 + 2689\,x^6 - 11520\,x^3$$

```
> expand(y^4 - 2*y^2 + 1);
```
$$x^8 + 12\,x^7 + 38\,x^6 - 36\,x^5 - 257\,x^4 + 132\,x^3 + 606\,x^2 - 720\,x + 225$$

Figure A.1: $\boxed{\text{A tutorial for beginners}}$

The up and down arrow keys $\boxed{\uparrow}$ and $\boxed{\downarrow}$ move to other lines, whereas the left and right arrow keys $\boxed{\leftarrow}$ and $\boxed{\rightarrow}$ move within a line. Any line can be altered and re-executed.

> *Warning:*   *Editing and re-executing commands in different lines can produce a worksheet that is very misleading to read.*

Using the mouse or arrow keys, return to the bottom-most prompt. Type `solve('')`; then press $\boxed{enter}$. You will see that Maple gives the two solutions to the equation $x^2 + 3x - 4 = 0$ in line 3 of Figure A.1. When you ask Maple to `solve` an expression that is not an equation, it attempts to find the roots; that is, Maple sets the expression equal to zero, and then solves.

Suppose that you want to evaluate the quadratic $x^2 + 3x - 4$ at $x = 2$. Use the substitution command `subs`. Type `subs(x=2, x^2+3*x-4);` at the prompt and then $\boxed{enter}$. The resulting value, 6, appears in line 4.

> *Note:*   *From this point on, we will not remind you to type the semicolon and then to press $\boxed{enter}$.*

In mathematics, an equal sign may denote an equation, such as $x^2 + 3x - 4 = 0$, or a logical statement, which might be true or false, such as $4 = 2 + 2$, or a definition, such as $a = 2$. The meaning intended is determined from the context, but there still is ambiguity in many instances. Maple uses the equal sign for the first two cases, but for the third, the definition, it is necessary to use "colon equal." Thus, to assign a the value 2, you would enter `a := 2`. Maple now knows that `a` represents the number 2. If you enter `a^2`, Maple will return the value 4 as in line 5 of Figure A.1. If you define b in terms of a, Maple will use 2 for each a, as in `b := 2*a - a^3` in line 6 of Figure A.1. The assignment will continue in force until you specifically change it or end the session. To clear or "undefine" a, enter `a := 'a'`. The effect is that the name a is assigned the letter a, thus removing any existing definition. Entering `restart` clears all definitions and begins a fresh session.

The "colon equal" can also be used to define functions. See Section A.4 for more details.

The easiest way to refer to an expression is by naming it. If you want the sixth power of the quadratic $x^2 + 3x - 4$, first define y to be the expression with `y := x^2+3*x-4`. Now enter `y^6` and $(x^2 + 3x - 4)^6$ appears. (See line 8 of Figure A.1.) In this way, complex expressions may be defined piece by piece and then assembled by referring to names. To expand y^6, use `expand`. Notice that the result in line 8 is so long that it "wraps around" the screen. Expand $y^4 - 2y^2 + 1$.

A.3   Help

FL: 1.06   FM: pg 3

Maple has two systems for getting information about commands and statements. The first is the *Help Browser*. Choosing Help from the menus opens a window with several panes. (See Figure A.2.) Each pane to the right carries subtopics of the topic selected in the pane to the left. The bottom pane shows a synopsis of the command. Clicking the Help button will open a window with the syntax, a detailed explanation, and examples of the selected command. Selecting an example and pressing $\boxed{enter}$ copies the example to the worksheet and executes it.

The second method of obtaining help is to enter a question mark followed by a topic or by a command name, and then press $\boxed{enter}$. The same detailed window is shown. Try ? `intro`. If Maple doesn't find the requested topic, several alternatives are offered.

Four Maple commands, new in Release 3, give brief information. The new commands are `info(`*name*`)`, `usage(`*name*`)`, `example(`*name*`)`, and `related(`*name*`)`. Enter ? `info` to see information for all four.

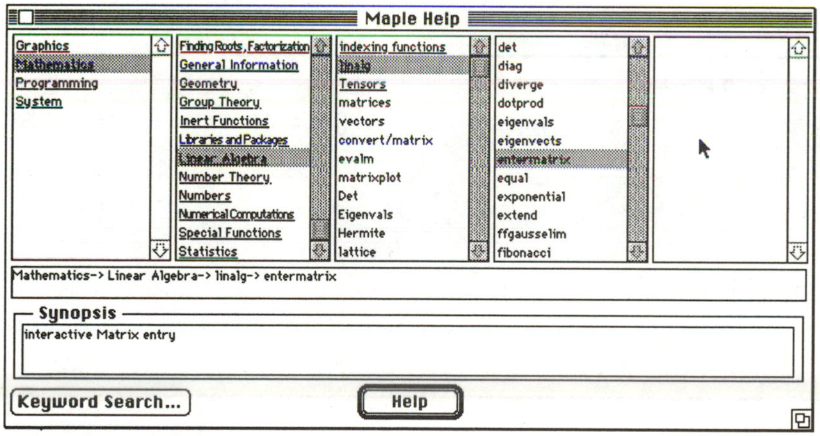

Figure A.2: $\boxed{\text{The Help Browser}}$

A.4 Defining Functions

FL: 1.18   FM: 2.3

There are many ways to define a function in Maple. We will look at the two easiest.

The first method uses the *arrow operator* and mimics the mathematical notation $f := (variables) \rightarrow (expression\ in\ variables)$. Since there is no arrow character on keyboards, Maple uses a diphthong of "-" and ">" to form the arrow. Maple's syntax is

$$name := variable(s) \rightarrow expression$$

as in `f := x -> x^2+3x-4`. To create a function of more than one variable, enclose the arguments in parentheses. For example, `g := (x,y) -> sin(x*y)` defines a function of two variables. After defining a function, we can use standard functional notation; for example, `f(3.1), f(Pi)`. Define $f := x \rightarrow x^2+3x-4$ and then expand $\frac{f(a+h)-f(a)}{h}$. See Figure A.3. Now try substituting $h = 0$ into the result. What did you find?

Our second way of defining functions is used for turning an existing expression into a function with `unapply`. The syntax is $name :=$ `unapply`($expression, variable$). The statement `f := unapply( x^2 + 3x - 4, x)` defines the function $f := x \rightarrow x^2 + 3x - 4$.

```
> f := x -> x^2+3x-4;
```
$$f := x \rightarrow x^2 + 3x - 4$$

```
> f(2/3);
```
$$-\frac{14}{9}$$

```
> (f(a+h)-f(a))/h;
```
$$2a + h + 3$$

Figure A.3: | Defining functions |

A.5 Entering Matrices and Vectors

FL: 2.13   FM: 4.1 to 4.3

Maple has a collection of matrix functions in a *package* called `linalg`. Bring these into your session by entering `with(linalg)` as in line 1 of Figure A.4. Use a colon, the *silent terminator*, instead of the usual semicolon to indicate to Maple to not show the output of a comand. (Do it once with a semicolon to see a list of all the linear algebra functions in the `linalg` package.)

There are several ways to enter a matrix. The first is to have Maple prompt you for each element. Define `A` as an arbitrary 2 by 2 matrix with `A := matrix(2,2)`. Now issue the `entermatrix(A)` command. See Figure A.4. Maple will ask you for the entries one at a time. You must type each entry followed by the usual semicolon and $\boxed{enter}$. Enter the elements as 1, 2, 3, and 4 in order one at a time. Observe that the message "enter element $i, j >$" appears above the line you enter the (i, j)-th element on as you proceed as seen in Figure A.4. Once you have finished, Maple will display the matrix $\begin{bmatrix} 1 & 2 \\ 3 & 4 \end{bmatrix}$.

```
> with(linalg):
Warning:  new definition for  norm
Warning:  new definition for  trace
> A := matrix(2,2);
```
$$A := \operatorname{array}(1..2, 1..2, [\quad])$$
```
> entermatrix(A);
enter element 1,1 >
> 1;
enter element 1,2 >
> 2;
enter element 2,1 >
> 3;
enter element 2, 2 >
> 4;
```
$$\begin{bmatrix} 1 & 2 \\ 3 & 4 \end{bmatrix}$$

Figure A.4: $\boxed{\text{Entering elements in a matrix, I}}$

The second method of entering matrices is quicker. The dimensions are typed first, followed by the elements in a list, all in one matrix statement. Enter `B := matrix(2,2, [1,2,3,4])` as in line 1 of Figure A.5. To see that `B` is the same as `A`, use the `linalg` package's `equal` command: `equal(A,B)`.

The next way to enter a matrix recognizes that, to Maple, a matrix is made from a list of lists. You can enter the matrix $\begin{bmatrix} 1 & 2 \\ 3 & 4 \end{bmatrix}$ by typing C := matrix([[1,2],[3,4]]) as in line 3 of Figure A.5. The rows are in lists separated by commas, with one set of brackets holding all row lists in one large list. Test C with equal(B,C).

```
> B := matrix(2,2, [1,2,3,4]);
```

$$\begin{bmatrix} 1 & 2 \\ 3 & 4 \end{bmatrix}$$

```
> equal(A,B);
```

$$true$$

```
> C := matrix([[1,2],[3,4]]);
```

$$\begin{bmatrix} 1 & 2 \\ 3 & 4 \end{bmatrix}$$

```
> equal(B,C);
```

$$true$$

Figure A.5: | Entering elements in a matrix, II |

The last technique makes use of a function that specifies how to calculate an entry given its indices. The *Hilbert* matrix is defined by $\mathbf{H} := [a_{i,j}]$ where $a_{ij} := 1 / (i+j-1)$. This matrix is more easily defined with a function than by listing the elements. Define h := (i,j) -> 1/(i+j-1), and then enter H := matrix(4,4, h). Maple returns a 4×4 Hilbert matrix. See Figure A.6.

In a similar fashion, to define a vector in Maple, we specify either a list of the elements or the size of the vector along with a function that generates the elements. To enter the vector $(1,2,3)$, type vector([1,2,3]) as in Figure A.6. Note the brackets around the list inside the parentheses. The generating function method can also be quite useful for vectors. Enter s := i -> i^2, and then enter u := vector(6,s) to generate the vector of squares $(1,4,9,16,25,36)$.

```
> h := (i,j) -> 1/(i+j-1);
```

$$h := (i,j) \rightarrow \frac{1}{i+j-1}$$

```
> H := matrix(4,4,h);
```

$$\begin{bmatrix} 1 & \frac{1}{2} & \frac{1}{3} & \frac{1}{4} \\ \frac{1}{2} & \frac{1}{3} & \frac{1}{4} & \frac{1}{5} \\ \frac{1}{3} & \frac{1}{4} & \frac{1}{5} & \frac{1}{6} \\ \frac{1}{4} & \frac{1}{5} & \frac{1}{6} & \frac{1}{7} \end{bmatrix}$$

```
> vector([1,2,3]);
```

$$\begin{bmatrix} 1 & 2 & 3 \end{bmatrix}$$

```
> s := i -> i^2:
  u := vector(6,s);
```

$$u := \begin{bmatrix} 1 & 4 & 9 & 16 & 25 & 36 \end{bmatrix}$$

Figure A.6: | Entering elements in a matrix, III |

What if I make a mistake typing?

If you find that you've made a mistake after the matrix appears on the screen and want to change a single entry, you can redefine it directly. For example, suppose the 2, 1-entry of A was mistakenly entered as 1 but should be 3. Just enter A[2,1] := 3 to fix the problem.

If you realize that you've made a mistake before finishing using entermatrix, you cannot back up. You must finish entering the elements and edit the result as we did above.

A.6   Uppercase and Lowercase Letters

FL: 1.9

Maple is case sensitive; that is, Maple distinguishes between upper- and lowercase letters. The names X and x are different, as are the names profit and Profit. Command names are also case sensitive; Matrix is not correct for matrix. The names of Maple's functions are "protected"; they can't be changed (easily). Try entering `sin := 1`.

Usually, it is convenient to assign a matrix a name such as $A := \begin{bmatrix} 1 & 2 \\ 3 & 4 \end{bmatrix}$. Since most books use uppercase letters for matrices, you may want to do so—we have.

A.7   Solving Systems of Equations

FL: 1.11   FM: 2.1

Systems of linear equations must appear in a set separated by commas and enclosed by set brackets or braces. If the system is homogeneous—that is, the constant terms are all zeros—then it is not necessary to type the "= 0". Asking Maple to solve either set of equations,

$$\{x + y, 2x - y\} \quad \text{or} \quad \{x + y = 0, 2x - y = 0\}$$

gives the same result.

When entering a large system of equations, it may be best to define each equation as a separate expression. Consider the following system of equations.

$$\begin{aligned} x + y + z &= 3 \\ x - y + 2z &= 4 \\ 2x - 3y - z &= 5 \end{aligned}$$

In Figure A.7, we have entered each equation separately as `eq1`, `eq2`, and `eq3`. The equations were assembled in line 2 using `sys := {eq1, eq2, eq3}`. Entering `solve{sys}` produces the result in line 3. The advantage of this method is that it avoids the need to type large expressions in one statement.

In linear algebra, we will often encounter the matrix form of a system such as

$$\begin{bmatrix} 1 & 2 \\ 3 & 4 \end{bmatrix} \begin{bmatrix} x \\ y \end{bmatrix} = \begin{bmatrix} 6 \\ 5 \end{bmatrix}$$

Unfortunately, `solve` does not handle matrix or vector equations directly. A matrix equation must be converted to a set of equations based on the entries in the matrices before being passed to `solve`. We will define our own function `matsolve` to deal with matrix/vector systems in a more efficient fashion. See Appendix B for further details.

```
>   eq1 := x + y + z = 3;
    eq2 := x - y + 2*z = 4;
    eq3 := 2*x - 3*y - z = 5;
```

$$eq1 := x + y + z = 3$$
$$eq2 := x - y + 2z = 4$$
$$eq3 := 2x - 3y - z = 5$$

```
> sys:={eq1,eq2,eq3};
```

$$\{x + y + z = 3, \quad x - y + 2z = 4, \quad 2x - 3y - z = 5\}$$

```
> solve(sys);
```

$$\left\{y = \frac{-2}{11}, \quad z = \frac{7}{11}, \quad x = \frac{28}{11}\right\}$$

Figure A.7: $\boxed{\text{Solving systems of equations system}}$

A.8 The Distinction Between Matrices and Vectors: An Important Warning

FM: 2.6 & 4.3

Most texts do not distinguish between the 1 by 3 matrix [1 2 3] and the vector [1, 2, 3]. Maple sees them as entirely different objects, and it is important to understand this to use the software. To see the difference, we'll enter the matrix [1 2] and the vector [1, 2]. Enter B := `matrix(1,2, [1,2])`. The result appears as line 1 of Figure A.8. Now enter the vector u := `vector([1,2])`. The result appears as line 2 of Figure A.8. If you examine the resulting displays carefully, you will see that the matrix in line 1 is contained in normal-style brackets, while the vector in line 2 is contained in bold-style brackets. This distinction in styles is very easy to miss.

The general idea is that a matrix is a vector whose entries are vectors of equal dimensions. Here's an important example: If you enter `transpose(A)`, Maple will return the transpose of the matrix as in line 4 of Figure A.8. If, on the other hand, you enter `transpose(u)`, Maple will simply return transpose (u) unevaluated as in line 5. To transpose a matrix is to switch its rows and columns. A vector does not have both rows *and* columns that can be switched.

```
> B := matrix(1,2, [1,2]);;
```
$$B := [1 \quad 2]$$

```
> u := vector([1,2]);
```
$$[1 \quad 2]$$

```
> convert(u, matrix);
```
$$\begin{bmatrix} 1 \\ 2 \end{bmatrix}$$

```
> transpose(B);
```
$$\begin{bmatrix} 1 \\ 2 \end{bmatrix}$$

```
> transpose(u);
```
$$\text{transpose}\,(u)$$

Figure A.8: | Maple's distinction between vectors and matrices |

A.9   Vector and Matrix Operations

FL: 2.13   FM: 4.1, 4.2, 4.3

To have Maple fully evaluate most matrix expressions, we must use **evalm**. That is, if we enter **A - 4*B**, Maple echoes $A - 4B$. To fully calculate, we must apply **evalm** to the expression either with **evalm('')** or originally with **evalm(A - 4*B)**.

Below we have collected the standard arithmetic operations and their syntaxes.

1. Maple's syntax for matrix and vector addition (**+**) and subtraction (**-**) are customary.

2. Taking a power of a square matrix uses (**^**), that is, **A^2**.

3. The dot product (or scalar product) of two vectors **u** and **v** is **dotprod(u,v)**. You may wish to shorten these long names; see Appendix B for methods.

4. Matrix multiplication is different from numeric multiplication. It is not commutative, and so has a different symbol. In Maple, matrix multiplication is denoted by &*, as in A &* B.

   There are situations that require extra care. For example, consider the following: Suppose we want to multiply the vector $[1, 2]$ by the matrix $\begin{bmatrix} 1 & 2 \\ 3 & 4 \end{bmatrix}$. Many books will treat $[1, 2]$ as a 1 by 2 matrix and write either $\begin{bmatrix} 1 & 2 \\ 3 & 4 \end{bmatrix} [1, 2]^T$ or $\begin{bmatrix} 1 & 2 \\ 3 & 4 \end{bmatrix} \begin{bmatrix} 1 \\ 2 \end{bmatrix}$. The way you indicate this product in Maple depends on whether you are dealing with the matrix [1 2] or the vector, $[1, 2]$. (The &* is required in either case.) In line 2 of Figure A.9, Maple cannot calculate the product because the matrices are not appropriately sized. In line 3, the product of the two matrices is correctly calculated using the transpose. In line 4 the product of the matrix and the vector is correctly calculated. (Notice that the output here is a vector while the output in line 3 is a matrix.) In line 5 the product is not calculated because the vector $[1, 2]$ does not have a transpose.

5. The three elementary row operations are defined in the linalg package.

   (a) mulrow(A,i,s) multiplies row i of matrix A by the number s.

   (b) addrow(A,i,j,s) adds s times row i of matrix A to row j.

   (c) swaprow(A,i,j) interchanges rows i and j of matrix A.

6. Matrix inversion is designated by raising to the power -1; e.g., A^(-1). The linalg package also contains the inverse function for inverting matrices.

7. The cross product of two vectors is denoted by crossprod(u,v).

8. The Euclidean length or norm (also called the 2-norm) of a vector is denoted by norm(u,2).

9. The transpose of a matrix A is denoted by transpose(A).

   Warning: Since vectors have only one dimension, a vector does not have a transpose in Maple.

10. The determinant of a matrix A is denoted by det(A).

11. The characteristic polynomial of a matrix A is denoted by `charpoly(A,x)`.

12. The eigen-system (eigenvalues with their multiplicities and corresponding eigen-vectors) is given by `eigenvects(A)`.

```
> A := matrix(2,2, [1,2,3,4]);
  B := matrix(1,2, [1,2]);
  u := vector([1,2]);
```

$$A := \begin{bmatrix} 1 & 2 \\ 3 & 4 \end{bmatrix}$$
$$B := \begin{bmatrix} 1 & 2 \end{bmatrix}$$
$$u := \begin{bmatrix} 1 & 2 \end{bmatrix}$$

```
> evalm(A &* B);      # an error returned
Error, (in linalg[multiply])
matrix dimensions incompatible
> evalm(A &* transpose(B));      # a matrix returned
```

$$\begin{bmatrix} 5 \\ 11 \end{bmatrix}$$

```
> evalm(A &* u);      # a vector returned
```

$$\begin{bmatrix} 1 & 2 \end{bmatrix}$$

```
> evalm(A &* transpose(u));      # an error returned
Error, (in linalg[multiply])
expecting a matrix or a vector
```

Figure A.9: Multiplication of matrices and vectors

A.10   Operating on the Elements of a Matrix or Vector

FL: 2.12   FM: 4.3

At times we need to apply a function to each element of a matrix or vector. The `map` function applies a specified operation to each element. For example, to take the natural logarithm of each element of a matrix A, we would `map(ln,A)`. If a function takes additional arguments, they follow the matrix. To find the (positive) residue mod 3 of the elements of the matrix $A := \begin{bmatrix} 1 & 2 \\ 3 & 4 \end{bmatrix}$, first define A with a matrix statement, and then `map(modp, A, 3)`. If we wished to simplify the elements of a vector u, we would use `map(simplify, u)`.

A.11 Solving Systems Approximately

FL: 1.12.2   FM: 2.1

The solutions to many systems may be very complex to express. It may be sufficient to know the solution approximately. Often the time involved in solving a system may be prohibitive; again an approximate solution may be found quickly. Maple's function for approximating solutions is `fsolve`. We show three general forms for `fsolve`

> `fsolve(`*equation, variable*`)`
> `fsolve(`*equation, variable=range*`)`
> `fsolve(`*equation, variable=range,*`complex)`

The second is for limiting the interval searched for roots. One disadvantage of `fsolve` is that, for nonlinear equations, usually only one root (if any) is returned; hence, it is important to specify the search interval. Applying `fsolve` to a polynomial equation with the third argument `complex` returns all the roots. See Figure A.10 for an example. Like `solve`, the `fsolve` function can also be applied to systems of equations. For further options, see `?fsolve`.

A.12 Graphing

FL: 0.1, 1.13, Chapter 4   FM: 2.3, 2.4, 5.2

One of Maple's strengths is its richness of plotting facilities. The simplest form is `plot(`*name*`)` where *name* is a function, as in `plot(cos)`. (See Figure A.10.) To specify the independent variables domain in this form, use `plot(`*name, start..finish*`)`; for example, to plot the square root function over the interval $[0, 5]$, you would enter `plot(sqrt, 0..5)`. If an expression is desired instead of a function name, the syntax is altered. We must now specify the independent variable with a range. To graph $x^2 - 2$ over the interval $[-2, 2]$, use `plot(x^2-2,x=-2..2)`. The dependent variables range can also be specified, along with the number of points tested, the line style, color, and so forth. See `?plot,options` for more information.

```
> y := x^5 + 5*x^2 - 1;
```

$$y := x^5 + 5x^2 - 1$$

```
> fsolve(y);
```

$$-1.667977977, \ -.4513841740, \ .4433661724$$

```
> fsolve(y, x, complex);
```

$$-1.667977977, \ -.4513841740, \ .4433661724, \ .8379979894 - 1.514422832\,I,$$
$$.8379979894 + 1.514422832\,I$$

```
> plot(y, x=-2..2, -10..10);
```

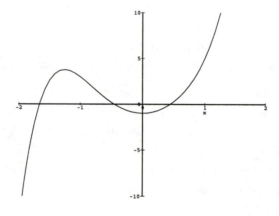

Figure A.10: $\boxed{\text{Approximate solutions}}$

Three-dimensional plotting is as easy. Try

$$\texttt{plot(sin(x*y)*(x\textasciicircum 2-y\textasciicircum 3), x=-1..1, y=-1..1)}$$

to see an interesting surface. Enter **?plot3d** and **?plot3d,options** for examples and information.

Maple's **plots** package has routines for graphing spacecurves, parametric surfaces, gradient fields, and 2- and 3-D animation, among other functions. See **?plots**.

A.13 Complex Numbers

FM: pg 107

Maple actually works in the field of complex numbers. Therefore, any of the examples in this book may be modified to involve complex numbers. The main thing to keep in mind is that the complex number i must be entered in Maple as I. Just as with matrices, we often need to apply the complex evaluator evalc to simplify complex expressions. It is often convenient to define a function to combine evaluation and simplification. Try

```
comp := z -> simplify(evalc(Re(z))) + simplify(evalc(Im(z)))*I
```

The functions Re and Im return the real and imaginary parts of z, respectively.

A.14 References

There is a growing library of material about Maple. Contact Waterloo Maple Software for the current list of publications and for information about *Maple Tech: The Maple Technical Newsletter*. Below is a collection of excellent resources

1. The Maple manuals are:

   - B. Char et al., *Maple V Library Reference Manual* (Springer-Verlag, 1991) ISBN 0-387-97592-6

   - B. Char et al., *Maple V Language Reference Manual* (Springer-Verlag, 1991) ISBN 0-387-97622-1

2. Two good introductions are:

   - B. Char et al., *First Leaves:A Tutorial Introduction to Maple V* (Springer-Verlag, 1992) ISBN 0-387-97621-3

   - W. Ellis et al., *Maple V Flight Manual* (Brooks/Cole Pub. Co., 1992) ISBN 0-534-21235-2

3. Two very useful, concise reference guides are:

   - N. Blachman and M. Mossingoff, *Maple V Quick Reference* (Brooks/Cole Pub. Co., 1994) ISBN 0-534-20478-3

- D. Redfern, *The Maple Handbook* (Springer-Verlag, 1994) ISBN 0-387-94054-5.

4. An excellent upper-level introduction is:

- A. Heck, *Introduction to Maple* (Springer-Verlag, 1993) ISBN 0-387-97662-0

5. For study of the theory of computer mathematics systems, the best choice is:

- K. Geddes et al., *Algorithms for Computer Algebra* (Kluwer Academic Pub., 1992) ISBN 0-792-39259-0

Appendix B
USER-DEFINED FUNCTIONS

There are several specialized tasks we need to do that Maple does not provide, hence we have to design the code ourselves. That is the purpose of this appendix.

Start with a new worksheet and follow the instructions carefully. Type everything *exactly* as it appears. The colon is essential in := and watch for the back-quote, ', which is usually located on the same key as the symbol $\sim$ on a standard keyboard. It is not an apostrophe (') or single quote. Also be careful to distinguish between brackets, [], and parentheses, ().

Loading files and doing things in the right order are important. Save your work using Maple's **save** statement followed by the file name. You may then use the functions in any Maple session by issuing a **read** command; for example, the command **read** 'LU.m' will read and define the LU function from the file named LU.m. The last line of each function worksheet below will save itself to a disk and restart Maple with a fresh session.

B.1 The `matsolve` function

<u>Discussion</u>: The Maple **solve** command, while extremely general and powerful, will not operate on matrix equations. By using Maple's **convert**, we can change a matrix equation into a set of equations that Maple will attempt to solve. To make the operation easier, we change the matrix equation $LHS = RHS$ into $LHS - RHS = 0$ before converting.

```
> matsolve := eq -> solve(convert(evalm(lhs(eq) - rhs(eq)), set));

> save(matsolve, 'matsolve.m');
    restart;
```

Figure B.1: `matsolve`

B.2 The `vlength` function

<u>Discussion</u>: Maple has a function for vector length, `norm`. However, `norm(v, p)` is a very general implementation and introduces absolute values as shown in the

definition $\|\mathbf{v}\|_p := \sqrt[p]{\sum_{i=1}^n |v_i|^p}$. The Euclidean, or 2-norm, does not need absolute values—the squares eliminate negatives. Since the absolute values in `norm` cause problems for Maple's `solve`, we define our length function using dot products; that is, $\|\mathbf{v}\| := \sqrt{\mathbf{v} \cdot \mathbf{v}}$.

```
> vlength := v -> sqrt(dotprod(v, v));

> save(vlength, 'vlength.m');
    restart;
```

Figure B.2: $\boxed{\texttt{vlength}}$

B.3   The `proj` function

<u>Discussion</u>: It is convenient to define a vector projection function to save typing in performing Gram-Schmidt calculations.

```
> proj := (u,v) -> innerprod(u, v)/innerprod(v, v) * v;

> save(proj, 'proj.m');
    restart;
```

Figure B.3: $\boxed{\texttt{proj}}$

B.4   The `LU` function

<u>Discussion</u>: The following code produces the *LU* decomposition of a matrix. See Chapter 18 for an example.

If *A* is a matrix that can be put into echelon form *without row interchanges*, then there is a pair of matrices, a lower triangular matrix *L* and an upper triangular matrix *U*, such that the following equation holds: *A = LU*.

This code is more complicated than most of the other examples in this appendix. It is designed to be as easy as possible to enter, at the expense of efficiency of calculation. Some of the lines of code such as `od` or `fi` may look like mistakes, but they are not. Begin with a clear screen and type each line exactly as it appears below.

If A is a matrix that can be put into echelon form *without row interchanges*, then LU(A) returns a lower triangular matrix (first element) and an upper triangular matrix (second element) in a *list* such that LU(A)[1] &* LU(A)[2] = A. This is the *LU* decomposition of A.

If a row interchange is encountered, the function ceases calculation and returns the error message "zero pivot."

```
> LU := proc(A:matrix) local n, AI, i, L, U;
  if det(A)=0 then
    ERROR('no decomposition: determinant is zero') fi;
  n := rank(A);
  AI := augment(A, band([1], n));
  for i from 1 to n-1 do
    if "[i,i]=0 then
      ERROR('no decomposition: zero pivot') fi;
    pivot(", i, i, (i+1)..n);
    mulrow(", i, 1/"[i,i]);
    od;
  L := delcols(", 1..n);
  U := delcols("", n+1..2*n);
  if det(L)<>0 then RETURN([inverse(L),eval(U)])
  else ERROR('error'); fi;
  end:

> save(LU, 'LU.m');
    restart;
```

Figure B.4: LU

B.5 The QR and QR_algorithm functions

<u>Discussion</u>: This code produces the QR factorization of A and executes the QR algorithm. See Chapter 18 for an example.

If A is a matrix whose columns are linearly independent, then there is a matrix Q with orthonormal columns and an invertible upper triangular matrix R such that

$A = QR$. This is the QR factorization of A. It should be noted that the matrix Q is precisely the result of applying the Gram-Schmidt process to the columns of A.

If A is a matrix whose columns are linearly independent, then `QR(A)` returns a list with the first element a matrix with orthonormal columns, and the second element an invertible upper triangular matrix such that `QR(A)[1] &* QR(A)[2] = A`.

The function `QR_algorithm` executes n iterations of the QR algorithm, returning a matrix that is similar to A and approximately upper triangular. See Chapter 18 for an example.

As with the LU program, begin with a fresh session and type each line exactly as it appears below.

```
> QR := proc(A:matrix) local vlist, i, Q, R;
  if rank(A)<>coldim(A) then
    ERROR('linearly dependent columns');
  else
    vlist := [seq(subvector(A, 1..rowdim(A), i), i=1..coldim(A))];
    vlist := GramSchmidt(vlist);
    vlist := map( v -> v/sqrt(dotprod(v, v)), vlist);
    Q := augment(op(vlist));
    R := inverse(Q) &* A;
    RETURN([evalm(Q), evalm(R)]);
    fi;
  end:

> QR_algorithm := proc(A:matrix, n:integer) local i, Ai, T;
  Ai := map(fnormal@evalf, copy(A));
  T := QR(Ai);
  Ai := map(fnormal, evalm(T[2] &* T[1]));
  if n>1 then QR_ algor(Ai, n-1);
  else evalm(Ai); fi:
  end:

> save(QR, QR_algorithm, 'QR.m');
    restart;
```

Figure B.5: | `QR` and `QR_algorithm` |

B.6   Selected Shorts

<u>Discussion</u>: There are many ways to tailor Maple to your taste. These can be used to shorten typing and input—both error-prone activities—or to use names that are easier to remember. Below we will offer just a few hints and shortcuts.

The `alias` command allows renaming objects with synonyms. Try a few.

```
alias(Id=&*())
alias(mat=evalm)
alias(apply=map)
```

A function can be turned into an operator by using Maple's *neutral operator* &. The backquotes are necessary.

```
'&.' := dotprod    makes  u &. v   the same as  dotprod(u,v)
'&x':= crossprod   makes  u &x v   the same as  crossprod(u,v)
'&m' := evalm      makes    &m A   the same as  evalm(A)
```

At times the `&*()` form of the identity matrix is not appropriate, for instance, when size is specified. An easy function to give I_n is

```
Id := n -> band([1],n)
```

In order to test conjectures, it is convenient to generate matrices and vectors with random elements. Maple's `randmatrix` and `randvector` functions help. The basic forms are

$$\texttt{randmatrix}(rows, columns)$$
$$\texttt{randvector}(length)$$

To extend these functions, Maple allows specifying how the entries are generated. For example, by using the random number generating function, we can have elements between, say, -2 and 5 by using

```
randmatrix(3,4, entries=rand(-2..5))
```

More complex functions can be made.

```
gen := () -> randpoly(x,degree=4)
randvector(4,entries=gen)
```

would produce a vector of length 4 that has random polynomials of degree 4.

Experiment!